Multiple Decremen

Shailaja Deshmukh

Multiple Decrement Models in Insurance

An Introduction Using R

 Springer

Shailaja Deshmukh
Department of Statistics
University of Pune
Pune, India

ISBN 978-81-322-1712-1   ISBN 978-81-322-0659-0 (eBook)
DOI 10.1007/978-81-322-0659-0
Springer New Delhi Heidelberg New York Dordrecht London

Springer is part of Springer Science+Business Media (www.springer.com)

Preface

Actuarial profession is one of the prestigious professions around the world. An actuary has to evaluate the entire operation of the insurance business deploying a variety of tools from actuarial science, which are heavily based on the statistical techniques and principles from finance and economics. With the liberalization of the insurance industry in India, the demand for actuaries and actuarial courses is increasing. The aim of this book is to cater to the needs of students intending to pursue actuarial profession and to familiarize them with application of various statistical methods used in the insurance industry. The book elaborates on actuarial concepts and statistical techniques in multiple decrement models with their application in pension funding, and multi-state transition models with application in disability income insurance. This book is written in the same style as my book "Actuarial Statistics: An Introduction Using R" published by Universities Press, India, in 2009, which discusses statistical tools for the computations of premiums and reserves for life insurance products and annuities in single decrement models, using R software.

In some policies, benefit to a single life or a group, is subject to a type of contingency. For example, the death of an individual may be due to an accident or due to any other cause. In most of the insurance policies, the coverage is firstly given for the base cause, and then there are policy riders for additional benefits. If death is due to an accident, then the benefit structure is different; usually benefit is more than the base coverage. In such cases, the benefit structure and consequently the premium structure depend on time to death and the cause of the death. Survivorship models incorporating two random mechanisms, time to termination and various modes of termination, are known as multiple decrement models. The first chapter introduces the multiple decrement model and the construction of the multiple decrement table using the associated single decrement model and central rate bridge.

Chapter 2 discusses calculation of premiums and reserves in life insurance products when the benefit depends on the cause of decrement along with the time to decrement.

A major application of multiple decrement models is in pension plans and employee benefit plans. In these schemes, the benefit paid on termination of employment depends upon the several causes of termination. The cause of termination may be withdrawal, disability, death, or retirement. The benefits on retirement often differ from those payable on death or disability. As a consequence, the actuarial present value of the benefits depends on the cause of death along with the future life time of an individual. To determine the rate of contribution in pension funds and to value the pension fund at specified times, it is necessary to find the actuarial present value of the benefits. Therefore, survivorship models for employee benefit systems and pension funds include random variables for both time to termination and cause of termination. Chapter 3 is devoted to the application of multiple decrement models for evaluating the cost of a given pension plan at a specific time. Once the estimate of the ultimate cost of the plan is determined, the next step is to determine the contributions required to pay for the estimated cost in an orderly manner, so that the estimated cost of the plan is spread over future years. These actuarial techniques are referred to as actuarial cost methods or actuarial funding methods. A funding method specifies the pattern, that is, the frequency, and the amount of aggregate contributions required to balance the benefit payments. Chapter 4 reviews some of these methods.

As an extension of multiple decrement models, the multi-state model of transitions is discussed in Chap. 5, when the transitions among the states are governed by Markov models. Multiple state models have proved to be appropriate models for an insurance policy in which the payment of benefits or premiums is dependent on being in a given state or moving between a given pair of states at a given time. Such a model is useful to decide premium in continuing care retirement communities model in health insurance and disability income insurance model in employee benefit schemes.

In all these chapters, it is assumed that the rate of interest in the calculations of actuarial present values is deterministic and usually constant over the period of policy. However, the assumption of deterministic interest will be rarely realized in practice, particularly for long-term policies. Chapter 6 introduces in brief stochastic models for interest rates and calculation of premiums for some products in this setup.

The highlight of the book is its usage of R software for statistical computations. R software is freely available from public domain. In all the Universities in India and abroad the use of R software is increasing tremendously. Most of the recent books incorporate R software for statistical analysis. To be consistent with the recent trend and demand, R software is used in this book to compute various monetary functions involved in insurance business. R commands are given for all the computations, and meanings of these are explained, so that a reader unfamiliar with R can also use it. All the tables inserted in the book and solutions to all illustrative examples are worked out using R. The command-driven R software brings out very clearly the successive stages in statistical computations.

The book builds on from the very basic concepts, defining and explaining the terms and to move on to their applications and actual computations with R. It is easy to follow and moves on step-by-step from basics to detailed calculations. The book contains many solved examples illustrating the theory. For better assimilation of the material, exercises are given at the end of each chapter. Statistical prerequisites to the book are concepts and computation of premiums and reserves for some standard insurance products based on single decrement models.

I hope that this book will be instructive and will induce interest among the students about the actuarial profession. I am sure the book will be helpful for those who wish to prepare for examinations conducted by actuarial societies worldwide.

Feedback, in the form of suggestions and comments from colleagues, students, and all readers, is most welcome.

I thank all my friends, colleagues, and family members for encouragement and support received throughout this venture. I am indeed thankful to the students who opted for this course in the last couple of years. They provided me the incentive to study rigorously and to collect and set a variety of problems, all of which are helpful in writing this book.

Pune, India   S. Deshmukh

Biography

Shailaja Deshmukh is a Professor of Statistics at the University of Pune, India. Her areas of interest are inference in stochastic processes, applied probability and analysis of microarray data. She has authored three books—Microarray Data: Statistical Analysis Using R (jointly with Dr. Sudha Purohit), Statistics Using R (jointly with Dr. Sudha Purohit and Prof. Sharad Gore) and Actuarial Statistics: An Introduction Using R. She has a number of research publications in various peer-reviewed journals.

Contents

List of Figures

List of Tables

Chapter 1
Multiple Decrement Models

1.1 Introduction

Calculation of premiums which are acceptable to both the parties, an insured and
an insurer, for a variety of insurance products, is one of the important computations
in insurance business. It is usually referred to as the rating or pricing of insurance
products. In the simplest setup, the interest rate is assumed to be constant throughout the active period of the policy, and the benefit payable to individual is contingent
on the event of death. The benefit may be payable to a group, and then it is necessary to define when to pay the benefit. Commonly used two approaches are payment
on the first death in a group (joint life status) or at the last death in the group (last
survivor status). In both the situations, the future life time is the only underlying random variable. Calculation of premiums and corresponding reserves in such setups
is discussed in many books, for example, Borowiak (2003), Bowers et al. (1997),
Deshmukh (2009), Dickson et al. (2009), and Promislow (2006). In some situations,
benefit to a single life or a group, is subject to a type of contingency. For example, a
death of an individual may be due to accident or due to any other causes. In most of
the insurance policies the coverage is firstly given for the base cause. In whole life
insurance policy death is a base cause, and then there are policy riders for additional
benefits, that is, if death is due to accident, then the benefit structure is different;
usually benefit is more than the base coverage. Of course there is additional premium for such an extra benefit. In such cases, the benefit structure and consequently
the premium structure depend on time to death and the cause of the death. The termination from a given status, being alive in this context, is known as a decrement.
Survivorship models incorporating two random mechanisms, time to termination,
and various modes of termination are known as multiple decrement models.

A major application of multiple decrement models is in pension plans and employee benefit plans. In employee benefit and pension schemes, the benefit paid on
termination of employment depends upon the several causes of termination. The
cause of termination may be withdrawal, disability, death, or retirement. The benefits on retirement often differ from those payable on death or disability. Benefit on
retirement is the monthly life annuity depending on the service period and the skill

S. Deshmukh, *Multiple Decrement Models in Insurance*,
DOI 10.1007/978-81-322-0659-0_1, © Springer India 2012

of the individual. If death occurs before retirement age, a lump sum may be payable to the beneficiary. If an individual withdraws from the employment, then there can be a deferred pension, or the individual's accumulated contribution may be returned. In case of disability, there may be additional benefit till the individual recovers completely. As a consequence, the actuarial present value of the benefits depends on the cause of death along with the future life time of an individual. To determine the rate of contribution in pension fund and to value the pension fund at specified times, it is necessary to find the actuarial present value of the benefits. Therefore, survivorship models for employee benefit systems and pension funds include random variables for both time to termination and cause of termination.

Following are some more situations where benefit structure depends on the cause of the death.

1. Most individual life insurance premiums are paid until death or until the end of the premium paying period as specified in the policy. However, in practice, some policy holders stop payments after some time. Many insurance products provide payment of some benefit even if premiums stop before the end of the specified premium payment term. However, an insurer has to decide how much to pay when claim from such a customer arises. In such situations, time until termination and cause of termination are the two random variables of interest to decide the benefit to be paid.

2. In an individual life insurance, a person may cease to be an active insured by (i) dying, (ii) withdrawing, (iii) becoming disabled, or (iv) reaching the end of the coverage period. Thus there are different modes of exit from the group of active insureds. Here also time until termination and cause of termination affect the benefit structure.

3. Disability income insurance provides periodic payments to insureds who satisfy the definition of disability specified in the policy. In some cases, the amount of periodic payments depends on whether the disability was caused by illness or accident. For example, if the disability is due to accident, then there may be periodic payment of P_1 units, and if the disability is due to illness, then there may be periodic payment of P_2 units, till the individual survives or till the end of coverage period as specified in the policy. To find the purchasing price of such policies, it is necessary to find the actuarial present value of the periodic payments. The actuarial present value of such payments is governed by two random mechanisms, future life time of the individual and the type of disability.

Multiple decrement models are commonly encountered in industrial applications and in public heath insurance. For example, failure of a metal strip under test conditions can occur in a number of different ways, by cracking, bucking, shearing, etc. In public health applications, it is of interest to study the incidence rates for various diseases. So data are collected on the cause of death, along with the data on age at death. Death may have been caused by cardiovascular disease, cancer, accident, or any other cause. Thus, in both these applications, the time to failure and cause of failure are two variables of interest. In actuarial science, we use the terminology of mortality and survival function, instead of failure rate or hazard rate as used in industrial applications. In actuarial science the primary interest is to see how to incorporate such dual information while determining the premiums and the actuarial present value of benefit in pension funding. In Biostatistics, theory of multiple decrement models is known as the theory of competing risks. In Sect. 1.2, it will be clear why it is referred to as theory of competing risks.

The theory of life table when there is a single mode of exit can be extended to a more general theory of multiple decrement models involving the effect of several causes of decrement on a group of individuals. The life table when the cause of death is not taken into account will be referred to as a single decrement table. The basic underlying random variable in a single decrement table is the future life time random variable. Chapter 3 of Bowers et al. (1997) and Chap. 4 of Deshmukh (2009) have discussed in detail the construction of life table, now referred to as single decrement table. In this chapter, we discuss the generalization of theory of single decrement model to the multiple decrement model in which we consider a group of a large number of lives subject to several causes of decrement. The multiple decrement table is very similar to the single decrement table, except that the l_x column is reduced by several d_x's rather than one. These several d_x columns correspond to several causes of decrement. Thus, in the theory of multiple decrement models, one more random variable comes in a picture, and that is the cause of termination. The next section introduces this second random variable $J(x)$ and discusses the joint distribution of $J(x)$ and the future life random variable $T(x)$.

1.2 Time to Decrement and Cause of Decrement Random Variables

Suppose that $T(x) \equiv T$ is a continuous random variable denoting the time until death of (x) in a single decrement model or time until termination from a status, joint life and last survivor status, as defined for multiple lives or time until exit from a certain group, such as employment with a particular employer. There may be more than one cause for termination from a given status. Suppose that $J(x) \equiv J$ is a discrete random variable denoting the cause of decrement random variable. Some illustrations, related to applications discussed in the previous section, are given below:

1. In pension funding and employee benefit applications, the random variable J takes values 1, 2, 3, or 4 depending on whether termination or exit from the group is due to withdrawal, disability, death, or retirement, respectively.
2. In life insurance applications, J could be assigned values 1 and 2 depending on whether the insured dies or chooses to terminate payment of premiums. If J takes value 1, then benefit may be 1 unit, and if $J = 2$, then benefit may be the fraction of accumulated value of the premiums paid.
3. In public health applications, J could be 1, 2, 3, 4 depending on whether death was caused by cardiovascular disease, cancer, accident, or any other cause.

We now study the joint distribution of T and J and the related marginal and conditional distributions, as these are basic ingredients in premium calculation. In this setup, T is a continuous random variable, while J is a discrete random variable with values $1, 2, \ldots, m$. We cannot have the joint probability density function or joint probability mass function of these two random variables. But we can define the joint distribution function and hence can find joint probabilities of interest. Suppose that

$$h_x(j) = h(j) = P\left[J(x) = j\right], \quad j = 1, 2, \ldots, m,$$

denotes the probability mass function of J and $g(t)$ denotes the probability density function of T. Suppose that

$$f(t, j)dt = P\left[t < T \leq t + dt, J = j\right]$$

is the probability of decrement in $(t, t + dt)$ due to cause j. Then the probability of decrement in the interval $[a, b]$ due to cause j and the probability of decrement in the interval $[a, b]$ due to any cause are respectively given by

$$P[a \leq T \leq b, J = j] = \int_a^b f(s, j)\, ds \quad \text{and} \quad P[a \leq T \leq b] = \sum_{j=1}^m \int_a^b f(s, j)\, ds.$$

It is to be noted that $f(t, j)$ does not have interpretation of joint density as T is continuous and J is discrete. The probability mass function $h(j)$ and the probability density function $g(t)$ are related to $f(t, j)$ as follows:

$$h(j) = P[J = j] = \int_0^\infty f(s, j)\, ds \quad \text{and} \quad g(t) = \sum_{j=1}^m f(t, j).$$

The international actuarial notation for the functions of the joint distribution of the time-until-decrement random variable and cause of termination in the multiple decrement model are as follows. $_t q_x^{(j)}$ denotes the probability of decrement in $(x, x + t)$ due to cause j and is given by

$$_t q_x^{(j)} = P\left[T(x) \leq t, J(x) = j\right] = \int_0^t f(s, j)\, ds \quad \text{and}$$

$$_\infty q_x^{(j)} = h(j), \quad j = 1, 2, \ldots, m.$$

Suppose that the symbol (τ) indicates that a function refers to all causes. Then,

$$_t q_x^{(\tau)} = P[T \leq t] = G(t) = \int_0^t g(s)\, ds,$$

$$_t p_x^{(\tau)} = 1 - {_t q_x^{(\tau)}} = P[T > t] = \frac{S(x + t)}{S(x)} = e^{-\int_0^t \mu_{x+s}^{(\tau)}\, ds}, \quad \text{and}$$

$$\mu_x^{(\tau)}(t) = \frac{g(t)}{1 - G(t)} = -\frac{1}{{}_t p_x^{(\tau)}} \frac{d}{dt} {}_t p_x^{(\tau)} = \frac{\mu_{x+t}^{(\tau)} {}_t p_x^{(\tau)}}{{}_t p_x^{(\tau)}} = \mu_{x+t}^{(\tau)}.$$

In the above functions, $\mu_{x+s}^{(\tau)}$ denotes the force of decrement corresponding to the life-length random variable X at age $x + s$, where decrement can occur due to any one of the m causes. $\mu_x^{(\tau)}(t)$ denotes the force of decrement corresponding to the random variable $T(x)$ at t. As in single decrement model, the force of decrement of $T(x)$ at t is the same as the force of decrement of X at $x + t$. ${}_t q_x^{(\tau)}$ denotes the distribution function of $T(x)$ at t, it is chance of decrement of (x) in $(x, x+t)$ due to any cause, conditional on survival up to age x. ${}_t p_x^{(\tau)}$ denotes the survival function of $T(x)$ at t; it is the probability of survival of (x) up to $x + t$. The force of decrement due to cause j, conditional on survival of (x) to $x + t$ is defined as

$$\mu_x^{(j)}(t) = \frac{f(t, j)}{1 - G(t)} = \frac{f(t, j)}{{}_t p_x^{(\tau)}}.$$

Since $f(t, j)$ does not have interpretation of joint density, $\mu_x^{(j)}(t)$ also does not have interpretation of conditional joint density. But $f(t, j)dt, j = 1, \ldots, m, t \geq 0$, can be expressed as

$$\begin{aligned} f(t, j)dt &= P[t < T(x) \leq t + dt, J = j] \\ &= P[T > t]P[t < T(x) \leq t + dt, J = j | T > t] \\ &= {}_t p_x^{(\tau)} \mu_{x+t}^{(j)} dt. \end{aligned}$$

Thus, $f(t, j) = {}_t p_x^{(\tau)} \mu_{x+t}^{(j)}, j = 1, \ldots, m, t \geq 0$. Substituting this expression in $\mu_x^{(j)}(t)$, we get $\mu_x^{(j)}(t) = \mu_{x+t}^{(j)}$.

Intuitively it is clear that the total force of decrement is the addition of the individual forces. We prove it algebraically in the following:

Result 1.2.1 *The total force of decrement is the sum of the forces of decrement due to m causes.*

Proof By definition,

$$ {}_t q_x^{(\tau)} = \int_0^t g(s) \, ds = \int_0^t \sum_{j=1}^m f(s, j) \, ds = \sum_{j=1}^m \int_0^t f(s, j) \, ds = \sum_{j=1}^m {}_t q_x^{(j)}.$$

Further,

$$ {}_t q_x^{(j)} = \int_0^t f(s, j) \, ds \quad \Rightarrow \quad \frac{d}{dt} {}_t q_x^{(j)} = f(t, j).$$

Therefore,

$$\mu_x^{(\tau)}(t) = \frac{g(t)}{1 - G(t)} = \frac{1}{{}_t p_x^{(\tau)}} \frac{d}{dt} {}_t q_x^{(\tau)} = \frac{1}{{}_t p_x^{(\tau)}} \frac{d}{dt} \sum_{j=1}^{m} {}_t q_x^{(j)}$$

$$= \frac{1}{{}_t p_x^{(\tau)}} \sum_{j=1}^{m} f(t, j) = \sum_{j=1}^{m} \frac{f(t, j)}{{}_t p_x^{(\tau)}} = \sum_{j=1}^{m} \mu_x^{(j)}(t),$$

and the result is proved.   □

The conditional probability mass function of J given decrement at time t is given by

$$h(j | T = t) = \frac{f(t, j)}{g(t)} = \frac{{}_t p_x^{(\tau)} \mu_{x+t}^{(j)}}{{}_t p_x^{(\tau)} \mu_{x+t}^{(\tau)}} = \frac{\mu_{x+t}^{(j)}}{\mu_{x+t}^{(\tau)}} = \mu_{x+t}^{(j)} / \sum_{j=1}^{m} \mu_{x+t}^{(j)}.$$

With the definitions of $\mu_{x+t}^{(j)}$ and $\mu_{x+t}^{(\tau)}$, we have for $j = 1, 2, \ldots, m$,

$$f(t, j) = {}_t p_x^{(\tau)} \mu_{x+t}^{(j)}, \qquad h(j) = {}_\infty q_x^{(j)}, \qquad g(t) = \sum_{j=1}^{m} f(t, j) = {}_t p_x^{(\tau)} \mu_{x+t}^{(\tau)},$$

and

$${}_t q_x^{(j)} = \int_0^t f(s, j)\, ds = \int_0^t {}_s p_x^{(\tau)} \mu_{x+s}^{(j)}\, ds.$$

It is to be noted that the expression for ${}_t q_x^{(j)}$ is similar to that of ${}_t q_x$, with an exception of superscript j. Further, the probability ${}_t q_x^{(j)}$ of decrement between ages x to $x + t$ due to cause j depends on ${}_s p_x^{(\tau)}$, $0 \le s \le t$, and thus on all the component forces of decrement. Consequently, when the forces for decrements other than j are increased,

$${}_t p_x^{(\tau)} = 1 - {}_t q_x^{(\tau)} = 1 - \sum_{j=1}^{m} {}_t q_x^{(j)}$$

is reduced, and hence ${}_t q_x^{(j)}$ also gets reduced. In view of this phenomenon, multiple decrement theory is also known as the theory of competing risks in survival analysis. The following examples illustrate the functions defined above for multiple decrement model and their interrelations.

Example 1.2.1 A multiple decrement model with two causes of decrement is given below in terms of the forces of decrement:

$$\mu_{x+t}^{(1)} = 0.0005 t a_1^x, \quad t \ge 0, \qquad \mu_{x+t}^{(2)} = 0.001 t a_2^x, \quad t \ge 0, \quad a_1, a_2 \ge 0.$$

Fig. 1.1 Graphs of $_tp_x^{(\tau)}$ for $x = 30, 40, 50,$ and 60

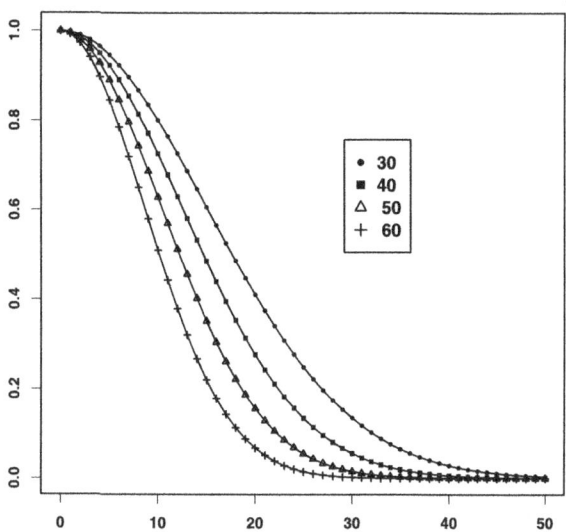

(i) For this model, find the expression for $_tp_x^{(\tau)}$, $f(t, j)$, $_tq_x^{(j)}$, $j = 1, 2$, and $g(t)$. Plot $_tp_x^{(\tau)}$, $f(t, j)$, $j = 1, 2$, and $g(t)$ for $x = 30, 40, 50,$ and 60 when $a_1 = 1.03$ and $a_2 = 1.04$.

(ii) Obtain the marginal distribution of J and conditional distribution of J given T.

Solution For the given two decrement model,

$$\mu_{x+s}^{(\tau)} = \mu_{x+s}^{(1)} + \mu_{x+s}^{(2)} = 0.0005sa_1^x + 0.001sa_2^x = 0.0005s\left(a_1^x + 2a_2^x\right).$$

Hence, the survival probability $_tp_x^{(\tau)}$ is given by

$$_tp_x^{(\tau)} = \exp\left[-\int_0^t \left[0.0005s\left(a_1^x + 2a_2^x\right)\right] ds\right]$$

$$= \exp\left[-0.00025t^2\left(a_1^x + 2a_2^x\right)\right], \quad t \geq 0.$$

Figure 1.1 shows the graph of $_tp_x^{(\tau)}$ for four ages 30, 40, 50, and 60. For all ages, $_tp_x^{(\tau)}$ is a decreasing function, being a survival function, and decrease is steep for age 60, as expected. The values of $_tp_{60}^{(\tau)}$ are almost 0 beyond $t = 30$.

The joint distribution of T and J is specified by

$$f(t, j) = \begin{cases} \exp[-0.00025t^2(a_1^x + 2a_2^x)]0.0005ta_1^x & \text{if } t \geq 0, \ j = 1, \\ \exp[-0.00025t^2(a_1^x + 2a_2^x)]0.001ta_2^x & \text{if } t \geq 0, \ j = 2. \end{cases}$$

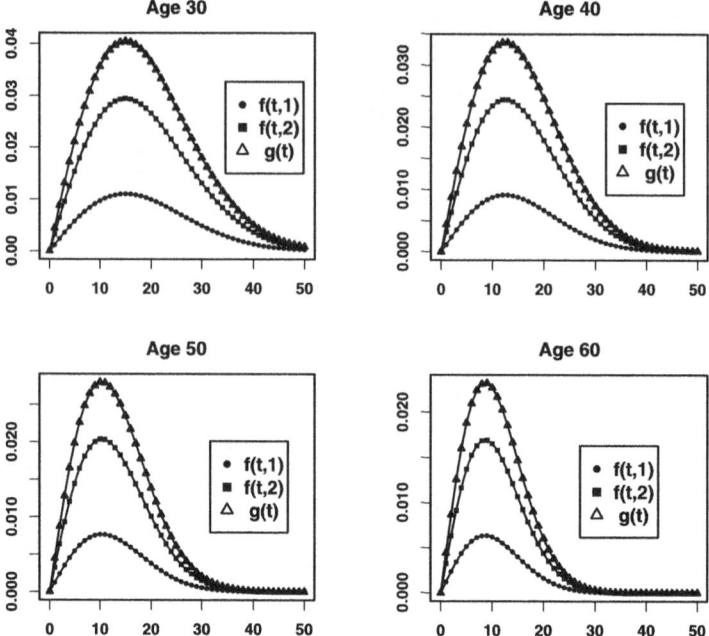

Fig. 1.2 Graphs of $g(t)$, $f(t, 1)$, and $f(t, 2)$

The marginal probability density function of T for $t \geq 0$ is

$$g(t) = \sum_{j=1}^{2} f(t, j) = {}_t p_x^{(\tau)} \mu_{x+t}^{(\tau)}$$

$$= \exp\left[-0.00025t^2\left(a_1^x + 2a_2^x\right)\right]\left[0.0005t\left(a_1^x + 2a_2^x\right)\right].$$

Figure 1.2 shows the graphs of $g(t)$, $f(t, 1)$, and $f(t, 2)$ for four ages 30, 40, 50, and 60. We note that shapes of the curves for $g(t)$, $f(t, 1)$, and $f(t, 2)$ for all ages are the same. However, as age increases, skewness increases.

By definition,

$$_tq_x^{(1)} = \int_0^t f(s, 1)\, ds = \frac{a_1^x[1 - e^{-(a_1^x + 2a_2^x)(0.00025t^2)}]}{a_1^x + 2a_2^x}$$

and

$$_tq_x^{(2)} = \int_0^t f(s, 2)\, ds = \frac{2a_2^x[1 - e^{-(a_1^x + 2a_2^x)(0.00025t^2)}]}{a_1^x + 2a_2^x}.$$

If we allow t to tend to ∞ in these two expressions, we get the marginal probability mass function of J. It is as follows:

$$h(j) = \begin{cases} \int_0^\infty f(t, 1) dt = \dfrac{a_1^x}{a_1^x + 2a_2^x} & \text{if } j = 1, \\[2mm] \int_0^\infty f(t, 2) dt = \dfrac{2a_2^x}{a_1^x + 2a_2^x} & \text{if } j = 2. \end{cases}$$

Finally the conditional probability mass function of J, given decrement at t, is

$$h(1|t) = \frac{a_1^x}{a_1^x + 2a_2^x}, \quad t \geq 0, \quad \text{and} \quad h(2|t) = \frac{2a_2^x}{a_1^x + 2a_2^x}, \quad t \geq 0.$$

It is to be noted that in this model the conditional distribution of J is the same as the marginal distribution of J. Thus, for this model, T and J are independent random variables.

Example 1.2.2 A multiple decrement model with two causes of decrement is specified by the following two forces of decrement:

$$\mu_{x+t}^{(1)} = BC^{x+t}, \quad t \geq 0, \quad \mu_{x+t}^{(2)} = A, \quad t \geq 0, \quad A \geq 0, \ B \geq 0, \ C \geq 1.$$

For this model, calculate $_t p_x^{(\tau)}$, $g(t)$, $f(t, j)$, $_t q_x^{(j)}$, $j = 1, 2$, and the marginal and conditional distributions of J given T.

Solution For this model, $\mu_{x+s}^{(\tau)} = \mu_{x+s}^{(1)} + \mu_{x+s}^{(2)} = BC^{x+s} + A$. It is to be noted that $\mu_{x+s}^{(1)}$ is the force of mortality corresponding to Gompertz' law while $\mu_{x+s}^{(\tau)}$ is the force of mortality corresponding to Makeham's law. The survival probability $_t p_x^{(\tau)}$ is given by

$$_t p_x^{(\tau)} = \exp\left[-\int_0^t \left[A + BC^{x+s}\right] ds\right] = \exp\left[-\left(At + mC^x\left(C^t - 1\right)\right)\right],$$

where $m = B / \log_e C$. Hence, the marginal probability density function of T is

$$g(t) = {}_t p_x^{(\tau)} \mu_{x+t}^{(\tau)} = \exp\left[-\left(At + mC^x\left(C^t - 1\right)\right)\right]\left(A + BC^{x+t}\right) \quad \text{if } t \geq 0,$$

and the joint distribution of T and J is specified by

$$f(t, j) = \begin{cases} \exp[-(At + mC^x(C^t - 1))](BC^{x+t}) & \text{if } t \geq 0, \ j = 1, \\ \exp[-(At + mC^x(C^t - 1))](A) & \text{if } t \geq 0, \ j = 2. \end{cases}$$

By definition,

$$_t q_x^{(1)} = \int_0^t f(s, 1) ds = \int_0^t \exp\left[-\left(As + mC^x\left(C^s - 1\right)\right)\right]\left(BC^{x+s}\right) ds.$$

To obtain the expression for $_tq_x^{(1)}$, denote mC^x by α_x. Then $f(s,1)$ is expressible as

$$f(s,1) = \alpha_x e^{\alpha_x} \log C e^{-As - \alpha_x C^s} C^s \quad \text{for } s > 0.$$

Substituting $C^s = y$, we get $e^{-As} = y^{-A/\log C}$, $C^s \log C\, ds = dy$, and the range of integration is from 1 to C^t. Suppose $(-A/\log C) + 1 = \lambda$. For Makeham's model to be an appropriate model for human life length, A is usually very small, and C ranges from 1 to 1.12. For this set of A and C, λ is always positive. Then $_tq_x^{(1)}$ simplifies to

$$_tq_x^{(1)} = \alpha_x e^{\alpha_x} \int_1^{C^t} e^{-\alpha_x y} y^{\lambda-1}\, dy.$$

When $\lambda > 0$, with some norming constant, integral can be expressed in terms of incomplete gamma function. Thus,

$$_tq_x^{(1)} = e^{\alpha_x} \Gamma(\lambda) \alpha_x^{1-\lambda} P\left[1 \le W \le C^t\right],$$

where W follows the gamma distribution with shape parameter λ and scale parameter α_x and with the probability density function $f_W(u) = f(u)$ given by

$$f(u) = \frac{\alpha_x^\lambda e^{-\alpha_x u} u^{\lambda-1}}{\Gamma(\lambda)}, \quad 0 < u < \infty.$$

Given the values of parameters A, B, and C and age x, the probability $P[1 \le W \le C^t]$ can be obtained using any statistical software such as minitab, matlab, SYSTAT, or R. Once we get $_tq_x^{(1)}$, $_tq_x^{(2)}$ is obtained as $_tq_x^{(2)} = {}_tq_x^{(\tau)} - {}_tq_x^{(1)} = 1 - {}_tp_x^{(\tau)} - {}_tq_x^{(1)}$. With this expression for $_tq_x^{(1)}$, the marginal probability mass function of J is obtained immediately by allowing t to tend to ∞ in $_tq_x^{(1)}$. Thus,

$$h(1) = \int_0^\infty f(t,1)\, dt = e^{\alpha_x} \Gamma(\lambda) \alpha_x^{1-\lambda} P[W \ge 1] \quad \text{and} \quad h(2) = 1 - h(1).$$

Finally the conditional probability mass function of J, given decrement at t, is

$$h(1|t) = \frac{BC^{x+t}}{A + BC^{x+t}}, \quad t \ge 0, \quad \text{and} \quad h(2|t) = \frac{A}{A + BC^{x+t}}, \quad t \ge 0.$$

For this model conditional distribution of J given T is different from the marginal distribution of J. Thus, T and J are not independent random variables.

In Example 1.2.3, we find numerical values of all the functions derived in Example 1.2.2 for specific values of the parameters. We use R software to compute these functions. R is used for all the computations in this book. Hence, a brief introduction to R is given below.

Introduction to R   R is a system for statistical analysis and graphics created by Ross Ihaka and Robert Gentleman (1996). R is both a software and a programming language considered as dialect of language S created by AT & T laboratories. R is advocated for a variety of reasons, some of which are given below.

(i) R is a free software. It can be obtained from Comprehensive R Archive Network (CRAN), which may be reached from the R project web site at www.r-project.org. The files needed to install R are distributed from this site. The instructions for installation are also available at this site.

(ii) It has an excellent built-in-help system and good facilities for drawing graphs.

(iii) R is a computer programming language; hence for those who are familiar with programming language, it will be very easy to write programs with user-written functions for required computations.

R is an interpreted language, meaning that all commands typed on the key board are directly executed without building in the complete programme like other programming languages such as C, Pascal, etc. Furthermore, the syntax of R is very simple and intuitive. Variables, data, functions, and results are stored in the active memory of R in the form of objects. The data analysis in R proceeds as an interactive dialogue with the interpreter. As soon as we type command at the prompt (>) and press the enter key, the interpreter responds by executing the command. R language includes the usual arithmetic operations, such as + for addition, − for subtraction, ∗ for multiplication, / for division, and ^ for exponentiation, with usual hierarchy. R uses the assignment operator <− ("less than" sign followed by "minus" sign) to give an object its value. For example, the command

```
y <- 20;
```

assigns the value 20 to object y.

In computation of monetary functions discussed in this book, we use the following built-in functions given in R. The most useful R function for entering small data sets is the c function, that is, combine function. This function combines elements together and constructs a vector. For example, the command

```
y <- c(10, 20, 30, 40);
```

constructs a vector with four elements 10, 20, 30, and 40. A command, length(y) specifies the number of elements in a vector. When a series of observations is stored in R as a vector, the standard arithmetic functions and operators apply to vectors on an element-wise basis. For example, the command for division of two vectors and its output is given below:

```
c(10, 20, 30, 40)/c(5, 4, 3, 2)
[1]  2  5  10  20
```

It is to be noted that the division is done element-wise. Such an operation is useful in the calculation of premiums for n-year term and endowment insurance for various values of n in one command. Observe the following command and its output:

Table 1.1 The output,
formatted with the help of
Excel

	x	y
1	10	2
2	20	3
3	30	4
4	40	5

```
c(10, 20, 30, 40)/2
[1]  5  10  15  20
```

In this illustration, denominator 2 is repeated 4 times to construct a vector of
length 4, each element of which is 2. To combine various column vectors of the
same length in a matrix form or a tabular form, we use the data.frame function
from R. Tables presented in all the chapters are constructed using the data.frame
function. Suppose that we have two vectors: $x = (10, 20, 30, 40)$, $y = (2, 3, 4, 5)$.
The following R command creates two vectors x and y:

```
x <- c(10, 20, 30, 40);    y <- c(2, 3, 4, 5);
```

The command

```
z <- data.frame(x, y);
```

constructs a table consisting of three columns, first corresponds to row number,
second corresponds to x, and the third corresponds to y. Table 1.1 presents the
output, formatted with the help of Excel.

In most applications, data sets are large, for example, values of $q_x^{(j)}$ for number
ages x values and $j = 1, 2, \ldots, m$ modes of decrement. Suppose that the Excel file
consists of $m + 1$ columns of values of x and $q_x^{(j)}$ for $j = 1, 2, \ldots, m$ and we want
to import these data from Excel for the calculation of premiums. The procedure for
this consists of the following steps. First we save the file as tab delimited text file
with some name, qx.txt, say on local disk D and then close the file. Suppose that
the columns have headings as x and $q_x^{(j)}$, $j = 1, 2, \ldots, m$. In R console type the
command

```
z <- read.table("D:/qx.txt", header=T);
```

This command stores the file qx.txt in data object z. The first argument of this
function gives the path of the file, which should be enclosed in quotes. The second
argument header is logical. It is TRUE if the columns in the data set are named,
otherwise it is FALSE. The command

```
x <- z[, 1];
```

extracts all elements of the first column of matrix or table z, which is a column of x
values and assigns it to object x. Similarly, the command

```
q <- z[, 2];
```

extracts all elements of the second column of z, which is a column of $q_x^{(1)}$ values and assigns it to object q.

Following are some commands frequently used in the book. The command t <- seq(0, 60, 1) generates a sequence of numbers from 0 to 60 with increment of 1 unit. sum(t) gives sum of the elements of object t. round(y, 4) command rounds the values in object y to 4 decimals. The command gamma(p) produces the value of gamma function at argument p, while the command pgamma(x, shape, scale) evaluates the distribution function of gamma distribution with specified shape and scale parameters at x. In previous versions of R, the irrational number e is stored as exp(1), but in recent version R 2.13.1 e is stored as e. We will use all these and some more functions in the computation of various monetary functions in the following chapters. Explanation of each function is given in front of each command. To find out more functions and its usage, one can use built-in-help system of R. For example,

?c   ?read.table   ?plot

will display information on these functions, its usage with illustration. For more details about R and use of R in statistical analysis, reader may refer to the books Verzani (2005), Purohit et al. (2008) and Deshmukh and Purohit (2007).

In Example 1.2.3, we find numerical values of all the functions derived in Example 1.2.2 for specific values of the parameters. The corresponding R commands are given for all the computations.

Example 1.2.3 Suppose that in Example 1.2.2, $A = 0.0008$, $B = 0.00011$, and $C = 1.095$.

(i) Draw the graph of survival probability $_t p_x^{(\tau)}$ of $T(x)$ for $x = 30, 40, 50, 60$.
(ii) Draw the graph of probability density function $g(t)$ of $T(x)$ for $x = 30, 40, 50, 60$.
(iii) Prepare a table of $_t p_x^{(\tau)}$ for $t = 1, 2, \ldots, 10$ and for $x = 30, 40, 50, 60$.
(iv) Prepare a table of $_t q_x^{(1)}$ and $_t q_x^{(2)}$ for $t = 1, 2, \ldots, 10$ and for $x = 30, 40, 50, 60$.
(v) Find the distribution of $J(x)$ for $x = 30, 40, 50, 60$.
(vi) Find the conditional distribution of $J(x)$ given $T(x) = 10$ for $x = 30, 40, 50, 60$.

Solution In Example 1.2.2 we have obtained expressions for all the required functions. We use R software to compute the values of various functions and their plots. Following is a set of R commands to obtain the numerical values of these functions for specified values of the parameters:

```
a1 <- 0.0008   #parameter A in μ(2)x+t;

b  <- 0.00011  #parameter B in μ(1)x+t;

a  <- 1.095   #parameter C in μ(1)x+t;
m  <- b/log(a, base=exp(1));
p  <- (-a1/log(a, base=exp(1)))+1 #parameter λ as defined
                                   # in Example 1.2.2;
```

The argument, base = exp(1), in the function log can be ignored as it is default. The following set of commands computes $_tp_x^{(\tau)}, _tq_x^{(\tau)}, _tq_x^{(1)}, _tq_x^{(2)}, g(t), h(1), h(2)$ for $x = 30$, which are stored in objects p1, q1, q11, q12, g1, h11, and h12, respectively:

```
x <- 30;
j <- m*a^x;    #αx as defined in Example 1.2.2 for x=30;
t <- seq(0, 60, 1);
e <- exp(1);
p1 <- e^(j-a1*t-j*(a^t))    #₁px(τ);
q1 <- 1-p1;    #₁qx(τ);
q11 <- e^j*gamma(p)*(j^(1-p))*(pgamma(a^t, p, j)
       -pgamma(1, p, j))    #₁qx(1);
q12 <- q1-q11    #₁qx(2);
g1 <- p1*(a1+b*a^(x+t))    # g(t);
h11 <- e^j*gamma(p)*(j^(1-p))*(1-pgamma(1, p, j))    #h(1);
h12 <- 1-h11    #h(2);
```

Suppose that p2, q2, q21, q22, g2, h21, h22 denote $_tp_x^{(\tau)}, _tq_x^{(\tau)}, _tq_x^{(1)}, _tq_x^{(2)}, g(t), f(t, 1), f(t, 2), h(1), h(2)$, respectively, for $x = 40$. Suppose that p3, q3, q31, q32, g3, h31, h32 denote $_tp_x^{(\tau)}, _tq_x^{(\tau)}, _tq_x^{(1)}, _tq_x^{(2)}, g(t), f(t, 1), f(t, 2), h(1), h(2)$, respectively, for $x = 50$ and p4, q4, q41, q42, g4, h41, h42 denote $_tp_x^{(\tau)}, _tq_x^{(\tau)}, _tq_x^{(1)}, _tq_x^{(2)}, g(t), f(t, 1), f(t, 2), h(1), h(2)$, respectively, for $x = 60$. Then the following set of commands produces a table of $_tp_x^{(\tau)}$ for $t = 1$ to 10 and $x = 30, 40, 50,$ and 60:

```
y <- data.frame(p1, p2, p3, p4);
y1 <- round(y, 4);
y2 <- y1[2:11, ];
t <- 1:10;
y3 <- data.frame(t, y2);
y3    #Table 1.2 of ₁px(τ) for t=1 to 10 and x=30,40,50 and 60;
```

From Table 1.2 we note that for all the ages, $_tp_x^{(\tau)}$ decreases as t increases. Further, rate of decrease increases as age increases. The following set of commands draws a graph of $_tp_x^{(\tau)}$ for $x = 30, 40, 50,$ and 60:

```
par(mfrow=c(1, 1), font.axis=2, font.lab=2,
    cex.axis=1, cex.lab=1.5, font=2, lwd=2);
plot(t, p1, "o", pch = 20, cex=0.7, main=" ",
     xlab=" ", ylab=" ");
lines(t, p2, "o", pch = 15, cex=0.7, main=" ");
lines(t, p3, "o", pch = 24, cex=0.7, main=" ");
lines(t, p4, "o", pch = 3, cex=0.7, main=" ");
legend(locator(1), pch=c(20, 15, 24, 3),
       legend=c("30", "40", "50", "60"), cex=1.2);
```

Table 1.2 Survival
probability

t	$_tp_{30}^{(\tau)}$	$_tp_{40}^{(\tau)}$	$_tp_{50}^{(\tau)}$	$_tp_{60}^{(\tau)}$
1	0.9975	0.9949	0.9885	0.9729
2	0.9947	0.9894	0.9761	0.9441
3	0.9919	0.9834	0.9629	0.9137
4	0.9888	0.9771	0.9486	0.8815
5	0.9855	0.9702	0.9333	0.8477
6	0.9820	0.9628	0.9169	0.8122
7	0.9783	0.9549	0.8993	0.7751
8	0.9743	0.9463	0.8805	0.7364
9	0.9700	0.9371	0.8604	0.6964
10	0.9653	0.9272	0.8391	0.6550

Fig. 1.3 Graph of survival
function of $T(x)$

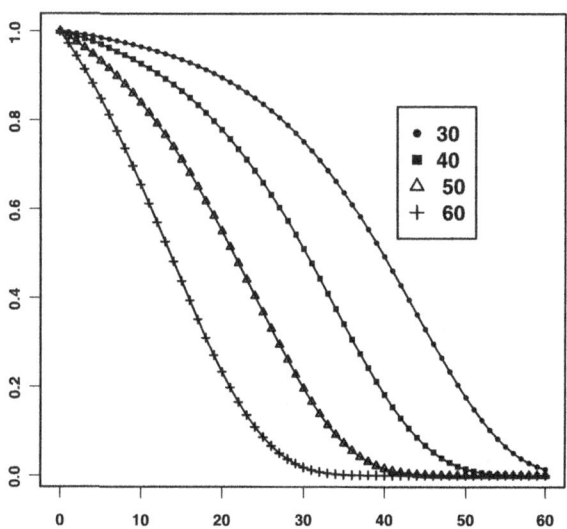

From Fig. 1.3 we note that for all ages, $_tp_x^{(\tau)}$ is a decreasing function, being a
survival function, and decrease is steep for age 60, as expected. The values of $_tp_{60}^{(\tau)}$
are almost 0 beyond $t = 30$. In the above set of commands which draws the graph
of $_tp_x^{(\tau)}$, if we replace p_1, p_2, p_3, p_4 by g_1, g_2, g_3, g_4, we get a set of commands to
draw a graph of $g(t)$ for $x = 30, 40, 50$, and 60 (see Fig. 1.4). As age increases, the
graph of $g(t)$ becomes more and more skew.

The following set of commands produces a table of $_tq_x^{(1)}$ and $_tq_x^{(2)}$ for $t = 1$ to
10 and $x = 30, 40, 50$, and 60:

```
d <- data.frame(q11, q12, q21, q22, q31, q32, q41, q42);
d1 <- round(d, 4);    d2 <- d1[2:11, ];
d3 <- data.frame(t, d2);
```

Fig. 1.4 Graph of probability
density function of $T(x)$

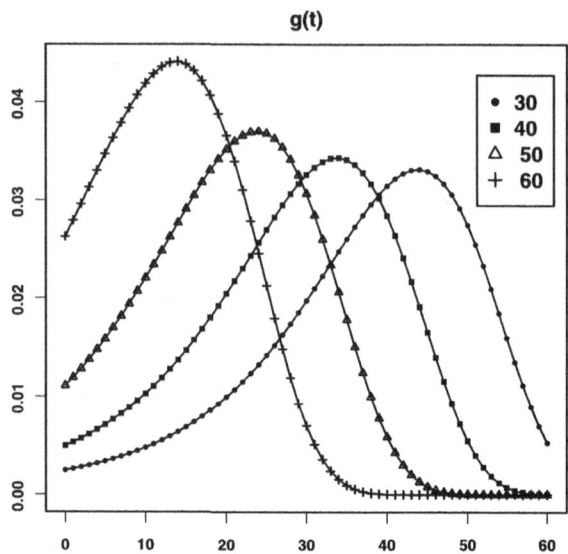

Table 1.3 Table of decrement probabilities

t	$_tq_{30}^{(1)}$	$_tq_{30}^{(2)}$	$_tq_{40}^{(1)}$	$_tq_{40}^{(2)}$	$_tq_{50}^{(1)}$	$_tq_{50}^{(2)}$	$_tq_{60}^{(1)}$	$_tq_{60}^{(2)}$
1	0.0018	0.0008	0.0043	0.0008	0.0107	0.0008	0.0263	0.0008
2	0.0037	0.0016	0.0091	0.0016	0.0223	0.0016	0.0543	0.0016
3	0.0057	0.0024	0.0142	0.0024	0.0348	0.0024	0.0840	0.0023
4	0.0080	0.0032	0.0198	0.0032	0.0483	0.0031	0.1154	0.0030
5	0.0105	0.0040	0.0259	0.0039	0.0629	0.0039	0.1486	0.0037
6	0.0132	0.0048	0.0325	0.0047	0.0785	0.0046	0.1834	0.0044
7	0.0162	0.0055	0.0396	0.0055	0.0954	0.0053	0.2199	0.0050
8	0.0194	0.0063	0.0474	0.0062	0.1135	0.0061	0.2580	0.0056
9	0.0229	0.0071	0.0559	0.0070	0.1328	0.0067	0.2975	0.0062
10	0.0268	0.0079	0.0651	0.0077	0.1535	0.0074	0.3382	0.0067

```
d3    #Table 1.3 of tqx(1) and tqx(2) for t=1 to 10 and
      #x = 30,40,50 and 60;
```

It is to be noted that $_tq_x^{(1)}$ increases as x increases and also as t increases, as expected. The values of $_tq_x^{(2)}$ are very small and do not vary much as age varies. The following set of commands produces probability mass function of $J(x)$ for $x = 30, 40, 50,$ and 60:

```
h1 <- c(h11, h21, h31, h41);
h2 <- c(h12, h22, h32, h42);
h <- round(data.frame(h1, h2), 4);
```

Table 1.4 Probability function of $J(x)$

Age	$h(1)$	$h(2)$
30	0.9697	0.0303
40	0.9768	0.0232
50	0.9833	0.0167
60	0.9889	0.0111

Table 1.5 Conditional distribution of $J(x)$ given $T(x) = 10$

x	$P[J(x) = 1|T(x) = 10]$	$P[J(x) = 2|T(x) = 10]$
30	0.8384	0.1616
40	0.9278	0.0722
50	0.9696	0.0304
60	0.9875	0.0125

```
age <- c(30, 40, 50, 60);
h3 <- data.frame(age, h);
h3    #Probability mass function of J(x) for x = 30, 40, 50,
      #and 60 as given in Table 1.4;
```

It is to be noted that chance of decrement due to nonaccident cause is higher than that for accident cause for all ages, which seems reasonable. Further, $h(1)$ increases with age, while $h(2)$ decreases with age. We find conditional distribution of $J(x)$ given $T(x) = 10$ for $x = 30, 40, 50$, and 60 using the following set of commands:

```
t <- 10;
x <- c(30, 40, 50, 60);
c11 <- b*a^(x+t)/(a1+b*a^(x+t));
c12 <- a1/(a1+b*a^(x+t));
d5 <- data.frame(c11, c12);
d6 <- round(d5, 4);
d8 <- data.frame(x, d6);
d8    #Conditional distribution of J(x) given T(x) = 10
      # for x = 30, 40, 50 and 60 as presented in Table 1.5;
```

In the conditional setup also, the probability of decrement due to cause 1 is higher than that for cause 2.

Example 1.2.4 A multiple decrement model has two causes of decrement. Both causes have constant forces of decrement $\mu^{(1)}$ and $\mu^{(2)}$. Suppose $_\infty q_x^{(1)} = 0.36$ and $E(T(x)) = 14$. Calculate $\mu^{(1)}$ and $\mu^{(2)}$.

Solution We know that

$$_\infty q_x^{(1)} = \int_0^\infty {_t p_x^{(\tau)}} \mu_{x+t}^{(1)} \, dt = \mu^{(1)} E(T(x)).$$

Hence,

$$\mu^{(1)} = \frac{{}_{\infty}q_x^{(1)}}{E(T(x))} = \frac{0.36}{14} = 0.0257 \quad \text{and}$$

$$\mu^{(2)} = \frac{{}_{\infty}q_x^{(2)}}{E(T(x))} = \frac{1 - {}_{\infty}q_x^{(1)}}{E(T(x))} = \frac{0.64}{14} = 0.0457.$$

We now define the discretized version of joint distribution of $T(x)$ and $J(x)$ analogous to the discretized version $K(x)$ of $T(x)$ defined for a single decrement model. It is useful to find the discrete premiums, that is, when benefit is payable at the end of year of death. In multiple decrement models, it has an application in pension funding also. In any pension plan there is a mandatory retirement age, an age at which all active employees must retire. In such instances, a continuous distribution for time T until termination is inadequate, as there is a positive probability of decrement at a specified time t and we have to work with the distribution of curtate-future time until decrement of (x). The random variable K, the curtate-future time until decrement of (x), is defined as the greatest integer strictly smaller than T, and it is similar to that in single decrement model. Using the joint distribution of T and J, we can write the joint probability mass function of K and J. Suppose that the possible values of J are $1, 2, \ldots, m$. Then the joint probability mass function of $K(x)$ and $J(x)$ is given by

$$P[K = k, J = j] = p(k, j) = P[k < T \le k+1, J = j]$$

$$= \int_k^{k+1} {}_t p_x^{(\tau)} \mu_{x+t}^{(j)} \, dt = \int_0^1 {}_{k+s} p_x^{(\tau)} \mu_{x+k+s}^{(j)} \, ds \quad (k+s = t)$$

$$= \int_0^1 {}_k p_x^{(\tau)} {}_s p_{x+k}^{(\tau)} \mu_{x+k+s}^{(j)} \, ds = {}_k p_x^{(\tau)} \int_0^1 {}_s p_{x+k}^{(\tau)} \mu_{x+k+s}^{(j)} \, ds$$

$$= {}_k p_x^{(\tau)} q_{x+k}^{(j)}.$$

Once we obtain the joint distribution of K and J, we get the marginal distributions and their expectations and variances. It is easy to see that

$$P[K(x) = k] = \sum_{j=1}^m p(k, j) = \sum_{j=1}^m {}_k p_x^{(\tau)} q_{x+k}^{(j)} = {}_k p_x^{(\tau)} q_{x+k}^{(\tau)} \quad \text{and}$$

$$P[J(x) = j] = h(j) = \sum_{k=0}^{\infty} p(k, j) = \sum_{k=0}^{\infty} P[k < T \le k+1, J(x) = j].$$

The conditional expectation of $K(x)$ given $J(x) = j$ is given by

$$E(K(x)|J(x) = j) = \sum_{k=0}^{\infty} k P[K(x) = k | J = j]$$

$$= \sum_{k=0}^{\infty} k p(k, j)/P[J = j], \quad j = 1, 2, \ldots, m.$$

The following example discusses the computations of all these functions for a two-decrement model, as specified in Example 1.2.2.

Example 1.2.5 A multiple decrement model with two causes of decrement is given below in terms of the forces of decrement as

$$\mu_{x+t}^{(1)} = BC^{x+t}, \quad t \geq 0, \qquad \mu_{x+t}^{(2)} = A, \quad t \geq 0, \quad A \geq 0, \ B \geq 0, \ C \geq 1.$$

For this model, with $A = 0.0008$, $B = 0.00011$, and $C = 1.095$, compute the joint probability mass function of $K(30)$ and $J(30)$. Find $E(K(30))$, $E(K(30)|J(30) = 1)$, and $E(K(30)|J(30) = 2)$.

Solution The joint probability mass function of $K(30)$ and $J(30)$ is given by

$$P\left[K(30) = k, J(30) = j\right] = {}_k p_{30}^{(\tau)} q_{30+k}^{(j)}.$$

In Example 1.2.2, we have obtained the expressions for ${}_t p_x^{(\tau)}$ and ${}_t q_x^{(1)}$. Further, ${}_t q_x^{(2)} = {}_t q_x^{(\tau)} - {}_t q_x^{(1)}$. Thus we have, ${}_k p_{30}^{(\tau)} = \exp[-Ak + \alpha_{30} - \alpha_{30} C^k]$ for $k = 0, 1, \ldots, 69$, where $\alpha_{30} = mC^{30}$. Here we assume that the limiting age is 100. It is valid for the given model with the given set of parameters as the probability of survival beyond 100 is 0.00002. $q_{x+k}^{(1)}$ is given by

$$q_{x+k}^{(1)} = e^{\alpha_{x+k}} \Gamma(\lambda) \alpha_{x+k}^{1-\lambda} P[1 \leq W \leq C],$$

where W follows the gamma distribution with shape parameter λ and scale parameter α_{x+k} and $\alpha_{x+k} = mC^{x+k}$. For the specified values of A, B, and C, λ is positive. Using these formulae for ${}_k p_{30}^{(\tau)}$ and $q_{30+k}^{(j)}$ for $j = 1, 2$, we have the following set of R commands to compute the joint mass function of $K(30)$ and $J(30)$:

```
a1 <- 0.0008   # A;
b <- 0.00011   # B;
a <- 1.095   # C;
m <- b/log(a, base=exp(1));
f <- (-a1/log(a, base=exp(1)))+1
                    # parameter λ as defined in Example 1.2.2;
x <- 30;
k <- 0:69;
j <- m*a^x;
j1 <- m*a^(x+k);
e <- exp(1);
p <- e^(-a1*k+j-j*a^k)   # vector of kp(τ)30 for k=0 to 69;
q1 <- e^j1*gamma(f)*(j1^(1-f))*(pgamma(a, f, j1)
```

```
          -pgamma(1, f, j1)) #vector of q^(1)_{30+k} for k=0 to 69;
q2 <- -e^(-a1-j1*a+j1)-q1 #vector of q^(2)_{30+k} for k=0 to69;
p1 <- p*q1    #vector of P[K(30)=k, J(30)=1] for k=0 to 69;
p2 <- p*q2    #vector of P[K(30)=k, J(30)=2] for k=0 to 69;
p3 <- p1+p2   #vector of P[K(30)=k] for k=0 to 69;
d <- round(data.frame(p1, p2, p3), 5);
d1 <- data.frame(k, d)   #Joint distribution of J(30) and
                         # K(30) and marginal distribution
                         #of K(30) as given in Table 1.6;
h1 <- sum(p1)   # P[J(30)=1];
h2 <- sum(p2)   # P[J(30)=2];
e30 <- sum(k*p3)   # E(K(30));
e130 <- sum(k*p1)/h1   # E(K(30)|J(30)=1);
e230 <- sum(k*p2)/h2   # E(K(30)|J(30)=2);
e30;   e130;   e230;
```

We get $h1 = P[J = 1] = 0.9697$, $h2 = P[J = 2] = 0.0303$, $E(K(30)) = 37.39$, $E(K(30)|J(30) = 1) = 37.91$, and $E(K(30)|J(30) = 2) = 20.66$. The joint probability mass function is presented in Table 1.6.

Here we have obtained $P[J(30) = 1] = 0.9697$ by summing $P[K(30) = k, J(30) = 1]$ over all values of k and $P[J(30) = 2] = 0.0303$ by summing $P[K(30) = k, J(30) = 2]$ for all k. These results are consistent with the results obtained in Example 1.2.3, using the definition $P[J(x) = j] = {}_\infty q_x^{(j)}$.

Example 1.2.6 Suppose that a multiple decrement model with two causes of decrement is as given in Example 1.2.1:

$$\mu_{x+t}^{(1)} = 0.0005ta_1^x, \quad t \geq 0, \qquad \mu_{x+t}^{(2)} = 0.001ta_2^x, \quad t \geq 0.$$

For this model, obtain the joint distribution of $K(x)$ and $J(x)$. Also obtain the marginal distribution of $K(x)$.

Solution The joint probability mass function of $K(x)$ and $J(x)$ is given by

$$P\big[K(x) = k, J(x) = j\big] = {}_k p_x^{(\tau)} q_{x+k}^{(j)}, \quad j = 1, 2.$$

For the given model, we have derived these functions in Example 1.2.1. Hence, for $k = 0, 1, \ldots,$

$$P\big[K(x) = k, J(x) = 1\big] = \frac{a_1^{x+k} e^{-0.00025k^2(a_1^x + 2a_2^x)}[1 - e^{-0.00025(a_1^{x+k} + 2a_2^{x+k})}]}{a_1^{x+k} + 2a_2^{x+k}}$$

and

$$P\big[K(x) = k, J(x) = 2\big] = \frac{2a_2^{x+k} e^{-0.00025k^2(a_1^x + 2a_2^x)}[1 - e^{-0.00025(a_1^{x+k} + 2a_2^{x+k})}]}{a_1^{x+k} + 2a_2^{x+k}}.$$

Table 1.6   Joint distribution of $K(30)$ and $J(30)$; Marginal distribution of $K(30)$

k	$p(k, 1)$	$p(k, 2)$	$p(k)$	k	$p(k, 1)$	$p(k, 2)$	$p(k)$
0	0.00175	0.00080	0.00255	35	0.02617	0.00050	0.02667
1	0.00191	0.00080	0.00271	36	0.02740	0.00048	0.02788
2	0.00209	0.00079	0.00288	37	0.02857	0.00045	0.02902
3	0.00228	0.00079	0.00307	38	0.02965	0.00043	0.03008
4	0.00249	0.00079	0.00328	39	0.03063	0.00041	0.03103
5	0.00271	0.00079	0.00350	40	0.03146	0.00038	0.03184
6	0.00296	0.00078	0.00375	41	0.03212	0.00036	0.03247
7	0.00323	0.00078	0.00401	42	0.03257	0.00033	0.03290
8	0.00352	0.00078	0.00430	43	0.03280	0.00030	0.03310
9	0.00384	0.00077	0.00461	44	0.03277	0.00028	0.03305
10	0.00419	0.00077	0.00495	45	0.03246	0.00025	0.03271
11	0.00455	0.00077	0.00532	46	0.03185	0.00022	0.03208
12	0.00496	0.00076	0.00572	47	0.03093	0.00020	0.03113
13	0.00540	0.00076	0.00615	48	0.02970	0.00017	0.02987
14	0.00587	0.00075	0.00662	49	0.02817	0.00015	0.02832
15	0.00638	0.00075	0.00713	50	0.02635	0.00013	0.02648
16	0.00693	0.00074	0.00767	51	0.02429	0.00011	0.02440
17	0.00752	0.00073	0.00826	52	0.02203	0.00009	0.02212
18	0.00816	0.00073	0.00889	53	0.01963	0.00007	0.01970
19	0.00885	0.00072	0.00957	54	0.01715	0.00006	0.01721
20	0.00958	0.00071	0.01029	55	0.01467	0.00005	0.01471
21	0.01036	0.00070	0.01107	56	0.01226	0.00003	0.01229
22	0.01120	0.00069	0.01189	57	0.00998	0.00003	0.01001
23	0.01209	0.00068	0.01277	58	0.00791	0.00002	0.00793
24	0.01303	0.00067	0.01371	59	0.00607	0.00001	0.00609
25	0.01403	0.00066	0.01469	60	0.00451	0.00001	0.00452
26	0.01508	0.00065	0.01573	61	0.00323	0.00001	0.00324
27	0.01618	0.00064	0.01682	62	0.00222	0.00000	0.00223
28	0.01733	0.00062	0.01796	63	0.00146	0.00000	0.00147
29	0.01853	0.00067	0.01914	64	0.00092	0.00000	0.00092
30	0.01976	0.00059	0.02035	65	0.00055	0.00000	0.00055
31	0.02102	0.00058	0.02160	66	0.00031	0.00000	0.00031
32	0.02231	0.00056	0.02287	67	0.00016	0.00000	0.00016
33	0.02361	0.00054	0.02415	68	0.00008	0.00000	0.00008
34	0.02490	0.00052	0.02542	69	0.00004	0.00000	0.00004

The marginal distribution of $K(x)$ is obtained by adding the above two expressions and is given by

$$P[K(x) = k] = e^{-0.00025k^2(a_1^x + 2a_2^x)}[1 - e^{-0.00025(a_1^{x+k} + 2a_2^{x+k})}].$$

For the specified values of a_1, a_2, and x, we use these expressions to compute the joint probability mass function of $K(x)$ and $J(x)$ and the marginal distribution of $K(x)$.

In the next section we will discuss how to construct the multiple decrement table using the multiple decrement model and the related functions defined in this section.

1.3 Multiple Decrement Table

Multiple decrement table is an extension of a single decrement table. In this setup the column of d_x is partitioned in m columns corresponding to m causes of decrement. We have noted in the previous section that once we have information on the force of decrement corresponding to m causes, we can find the survival probability and the decrement probabilities corresponding to all the causes. Thus, as in the setup of single decrement table, we first consider a random survivorship group approach of constructing a life table corresponding to m causes of decrement. Suppose that we have a group of $l_a^{(\tau)}$ lives, each of age a years. Each life is assumed to have the same joint distribution of time, until decrement and cause of decrement, specified by the joint probability

$$f(t, j) dt = {}_t p_x^{(\tau)} \mu_{x+t}^{(j)} dt, \quad t \geq 0, \ j = 1, \ldots, m.$$

Suppose that $\mathcal{L}_x^{(\tau)}$ is a random variable indicating the number of survivors at age x out of the $l_a^{(\tau)}$ lives in the original group at age a. Then $\mathcal{L}_x^{(\tau)}$ can be expressed as $\mathcal{L}_x^{(\tau)} = \sum_{i=1}^{l_a^{(\tau)}} Z_i$ where Z_i is defined as, $Z_i = 1$ if the ith life survives up to age x, $x \geq a$, and 0 otherwise. Then $E(Z_i) = P[Z_i = 1] = P[T(a) \geq x]$, the same for all i. Thus, expectation of $\mathcal{L}_x^{(\tau)}$, denoted by $l_x^{(\tau)}$, is given by

$$l_x^{(\tau)} = E(\mathcal{L}_x^{(\tau)}) = l_a^{(\tau)} P[T(a) \geq x] = l_a^{(\tau)} {}_{x-a} p_a^{(\tau)}.$$

Further, as in single decrement table, we get

$$l_{x+1}^{(\tau)} = l_a^{(\tau)} {}_{x+1-a} p_a^{(\tau)} = l_a^{(\tau)} {}_{x-a} p_a^{(\tau)} p_x^{(\tau)} = l_x^{(\tau)} p_x^{(\tau)}.$$

To obtain the analogue of ${}_n d_x$, for each such life, a Bernoulli random variable Y_j is defined as $Y_j = 1$ if an individual from original group of $l_a^{(\tau)}$ individuals suffers decrement in $(x, x+n)$, $x \geq a$, due to cause j and 0 otherwise. Then,

$$P[Y_j = 1] = P[x - a \leq T(a) \leq x + n - a, J(a) = j] = \int_{x-a}^{x+n-a} {}_t p_a^{(\tau)} \mu_{a+t}^{(j)} dt$$

$$= \int_0^n u+(x-a) p_a^{(\tau)} \mu_{u+x}^{(j)} \, du, \quad \text{with } t - (x - a) = u,$$

$$= \int_0^n x-a p_a^{(\tau)} \, _u p_x^{(\tau)} \mu_{u+x}^{(j)} \, du = x-a p_a^{(\tau)} \int_0^n {}_u p_x^{(\tau)} \mu_{u+x}^{(j)} \, du$$

$$= x-a p_a^{(\tau)} \, {}_n q_x^{(j)}.$$

Suppose that the random variable $_n \mathcal{D}_x^{(j)}$ denotes the number of lives who leave the group between ages x and $x + n$, $x \geq a$, from cause j. Then

$$_n \mathcal{D}_x^{(j)} = \sum_{i=1}^{l_a^{(\tau)}} Y_{ji}.$$

Its expectation is denoted by $_n d_x^{(j)}$ and is given by

$$_n d_x^{(j)} = E\left(_n \mathcal{D}_x^{(j)} \right) = l_a^{(\tau)} P[Y_j = 1] = l_a^{(\tau)} \, x-a p_a^{(\tau)} \, _n q_x^{(j)} = l_x^{(\tau)} \, _n q_x^{(j)}.$$

When $n = 1$, the prefixes are deleted. Thus, $d_x^{(j)} = l_x^{(\tau)} q_x^{(j)}$. Suppose that $_n \mathcal{D}_x^{(\tau)}$ denotes the number of decrements due to all causes between $(x, x + n)$, $x \geq a$. Then

$$_n \mathcal{D}_x^{(\tau)} = \sum_{j=1}^m {}_n \mathcal{D}_x^{(j)}.$$

Taking the expectations, we get

$$_n d_x^{(\tau)} = E\left(_n \mathcal{D}_x^{(\tau)} \right) = \sum_{j=1}^m {}_n d_x^{(j)} = \sum_{j=1}^m l_x^{(\tau)} \, _n q_x^{(j)} = l_x^{(\tau)} \, _n q_x^{(\tau)}.$$

With $n = 1$, $d_x^{(\tau)} = l_x^{(\tau)} q_x^{(\tau)}$. Further, we note that

$$l_{x+1}^{(\tau)} = l_x^{(\tau)} p_x^{(\tau)} = l_x^{(\tau)} \left[1 - \sum_{j=1}^m q_x^{(j)} \right] = l_x^{(\tau)} - \sum_{j=1}^m d_x^{(j)} = l_x^{(\tau)} - d_x^{(\tau)}.$$

These results enable us to obtain $l_x^{(\tau)}$ and $d_x^{(j)}$ values from $p_x^{(\tau)}$ and $q_x^{(j)}$ values. The table depicting the values of $p_x^{(\tau)}$ and $q_x^{(j)}$, $j = 1, \ldots, m$, or $l_x^{(\tau)}$ and $d_x^{(j)}$, $j = 1, \ldots, m$, and for integral values of x, is known as a multiple decrement table.

The functions $l_x^{(\tau)}$, $d_x^{(\tau)}$, $q_x^{(\tau)}$ defined above can be viewed from the other angle, and that is nothing but a deterministic survivorship group approach. Thus, the total force of decrement is viewed as a total rate of decrement. In this approach, a group of $l_a^{(\tau)}$ lives advances through age subject to deterministic forces of decrement $\mu_y^{(\tau)}$, $y \geq a$. The number of survivors to age x from the original group of $l_a^{(\tau)}$ lives at age

a is given by

$$l_x^{(\tau)} = l_a^{(\tau)} \exp\left[-\int_a^x \mu_y^{(\tau)}\, dy\right],$$

and the total decrement between ages x and $x+1$ is given by

$$d_x^{(\tau)} = l_x^{(\tau)} - l_{x+1}^{(\tau)} = l_x^{(\tau)}\left[1 - \frac{l_{x+1}^{(\tau)}}{l_x^{(\tau)}}\right]$$

$$= l_x^{(\tau)}\left[1 - \exp\left[-\int_x^{x+1} \mu_y^{(\tau)}\, dy\right]\right] = l_x^{(\tau)}\left(1 - p_x^{(\tau)}\right) = l_x^{(\tau)} q_x^{(\tau)}.$$

$q_x^{(\tau)}$ is interpreted as the effective annual total rate of decrement between the age x to $x+1$ governed by the forces $\mu_y^{(\tau)}$, $x \le y \le x+1$. By definition, $l_x^{(\tau)}$ is a differentiable function of x, when age x is treated as a continuous variable. Hence,

$$\frac{d}{dx}\, l_x^{(\tau)} = l_a^{(\tau)} \exp\left[-\int_a^x \mu_y^{(\tau)}\, dy\right]\left[-\mu_x^{(\tau)}\right] = -\mu_x^{(\tau)} l_a^{(\tau)}\,_{x-a}p_a^{(\tau)} = -\mu_x^{(\tau)} l_x^{(\tau)}$$

$$\Rightarrow \quad \mu_x^{(\tau)} = -\frac{1}{l_x^{(\tau)}} \frac{d}{dx}\, l_x^{(\tau)} = -\frac{d}{dx} \log l_x^{(\tau)}.$$

Thus, $\mu_x^{(\tau)}$ is interpreted as the rate of decrement of $\log l_x^{(\tau)}$.

With m causes of decrement, $l_x^{(\tau)}$ survivors will be classified into distinct sub-groups $l_x^{(j)}$, which denotes the number from the $l_x^{(\tau)}$ survivors who will suffer decrement in future ages due to cause j, so $l_x^{(\tau)} = \sum_{j=1}^m l_x^{(j)}$. In the deterministic approach, the force of decrement at age x due to cause j is defined as

$$\mu_x^{(j)} = \lim_{h \to 0} \frac{l_x^{(j)} - l_{x+h}^{(j)}}{h l_x^{(\tau)}} = -\frac{1}{l_x^{(\tau)}} \frac{d}{dx}\, l_x^{(j)}.$$

In this approach also, the total force of decrement is the sum of the forces of decrement due to various causes, as is clear from the following. We have

$$\mu_x^{(\tau)} = -\frac{1}{l_x^{(\tau)}} \frac{d}{dx} l_x^{(\tau)} = -\frac{1}{l_x^{(\tau)}} \frac{d}{dx} \sum_{j=1}^m l_x^{(j)} = -\frac{1}{l_x^{(\tau)}} \sum_{j=1}^m -\mu_x^{(j)} l_x^{(j)} = \sum_{j=1}^m \mu_x^{(j)}.$$

Let $q_x^{(j)}$ denote the proportion of the $l_x^{(\tau)}$ survivors to age x who terminate due to cause j before age $x+1$ when all m causes of decrement are operating. To obtain the expression for $q_x^{(j)}$ in the deterministic approach, recall that

$$\mu_y^{(j)} = -\frac{1}{l_y^{(\tau)}} \frac{d}{dy} l_y^{(j)} \quad \Rightarrow \quad -dl_y^{(j)} = \mu_y^{(j)} l_y^{(\tau)}\, dy.$$

Table 1.7 Decrement
probabilities

x	$q_x^{(1)}$	$q_x^{(2)}$
50	0.00490	0.01
51	0.00537	0.02
52	0.00590	0.03
53	0.00647	0.04
54	0.00708	0.05
55	0.00773	0.06
56	0.00844	0.06
57	0.00926	0.07
58	0.01019	0.08
59	0.01120	0.09

Therefore,

$$\int_x^{x+1} -dl_y^{(j)} = \int_x^{x+1} l_y^{(\tau)} \mu_y^{(j)}\, dy \quad \Leftrightarrow \quad l_x^{(j)} - l_{x+1}^{(j)} = d_x^{(j)} = \int_x^{x+1} l_y^{(\tau)} \mu_y^{(j)}\, dy.$$

Hence,

$$\frac{d_x^{(j)}}{l_x^{(\tau)}} = \int_x^{x+1} \frac{l_y^{(\tau)}}{l_x^{(\tau)}} \mu_y^{(j)}\, du = \int_x^{x+1} {}_{y-x}p_x^{(\tau)} \mu_y^{(j)}\, dy = \int_0^1 {}_u p_x^{(\tau)} \mu_{x+u}^{(j)}\, dy = q_x^{(j)}.$$

Thus, we get the same expression for $q_x^{(j)}$ as in the random survivorship approach.
As in the case of single decrement table, the deterministic approach provides an
alternative language and conceptual framework for multiple decrement theory. In
summary, given $q_x^{(j)}$ for $j = 1, 2, \ldots, m$ and $l_x^{(\tau)}$, we find for $x = 1, 2, \ldots,$

$$p_x^{(\tau)} = 1 - \sum_{j=1}^m q_x^{(j)}, \qquad l_{x+1}^{(\tau)} = l_x^{(\tau)} p_x^{(\tau)}, \qquad d_x^{(j)} = l_x^{(\tau)} q_x^{(j)}.$$

On the other hand, given $d_x^{(j)}$ for $j = 1, 2, \ldots, m$ and $l_x^{(\tau)}$, we find for $x = 1, 2, \ldots,$

$$q_x^{(j)} = d_x^{(j)} / l_x^{(\tau)}, \quad j = 1, 2, \ldots, n, \quad \text{and} \quad p_x^{(\tau)} = 1 - \sum_{j=1}^m q_x^{(j)}.$$

The following examples illustrate how the multiple decrement tables are constructed
in both these set ups.

Example 1.3.1 Table 1.7 gives the probability of decrement due to two causes.
Cause 1 is a death, and cause 2 is retirement. Age of mandatory retirement is 60

years. Suppose that there are 1000 individuals of age 50 working in a company and they are subject to the decrement according to the probabilities given in the table. Find the expected number of individuals who retire at 60. Also find the expected number of retirements and expected number of deaths in each of the year from 50 to 59.

Solution From the given data we note that $q_x^{(j)} > 0$ for $50 \leq x \leq 59$. Thus, it is implicitly assumed that the minimum eligible age for retirement is 50. The expected number of individuals who retire at 60 is l_{60} in a two-way decrement table, as 60 is the mandatory age of retirement. To find the expected number of retirements and expected number of deaths in each of the year from 50 to 59, we need to construct the two-decrement table. We construct the table using the formulae summarized above and the following R commands. Suppose that Table 1.7 is saved as a file m2.txt on drive D. We begin with importing the data file to R console:

```
z <- read.table("D://m2.txt", header=T);
x <- z[, 1];
q1 <- z[, 2]     # q_x^{(1)};
q2 <- z[, 3]     # q_x^{(2)};
q <- q1+q2     # q_x^{(τ)};
p <- 1-q     # p_x^{(τ)};
w <- length(p);
l1 <- 1000;
l <- c(l1, 2:w)    # dummy vector to store l_x^{(τ)};
for(i in 2:w)
   {
   l[i] <- l[i-1]*p[i-1];
   }
d1 <- l*q1    # d_x^{(1)};
d2 <- l*q2    # d_x^{(2)};
y <- data.frame(l, d1, d2);
y1 <- round(y, 2);
y2 <- data.frame(x, q1, q2, p, y1);
y2   # Table 1.8;
a <- l[w]-d1[w]-d2[w]    # l_{60};
a;
```

Thus, $l_{60}^{(\tau)} = l_{59}^{(\tau)} - d_{59}^{(1)} - d_{59}^{(2)} = 544.1952$, that is, out of a group of 1000 individuals of age 50, the expected number of individuals who retire at age 60 is 544. Column $d_x^{(2)}$ specifies the expected number of early retirements in age interval $(x, x+1)$ from ages 50 to 59. Column $d_x^{(1)}$ specifies the expected number of deaths between ages x to $x+1$, $x = 50, 51, \ldots, 59$.

Table 1.8 Expected number of decrements

Age x	$q_x^{(1)}$	$q_x^{(2)}$	$p_x^{(\tau)}$	$l_x^{(\tau)}$	$d_x^{(1)}$	$d_x^{(2)}$
50	0.00490	0.01	0.98510	1000.00	4.90	10.00
51	0.00537	0.02	0.97463	985.10	5.29	19.70
52	0.00590	0.03	0.9641	960.11	5.66	28.80
53	0.00647	0.04	0.95353	925.64	5.99	37.03
54	0.00708	0.05	0.94292	882.63	6.25	44.13
55	0.00773	0.06	0.93227	832.25	6.43	49.93
56	0.00844	0.06	0.93156	775.88	6.55	46.55
57	0.00926	0.07	0.92074	722.78	6.69	50.59
58	0.01019	0.08	0.90981	665.49	6.78	53.24
59	0.01120	0.09	0.89880	605.47	6.78	54.49

Table 1.9 Expected number of decrements

x	$l_x^{(\tau)}$	$d_x^{(1)}$	$d_x^{(2)}$
50	1000	10	15
51	975	11	16
52	948	12	16
53	920	13	17
54	890	13	18
55	859	15	20
56	824	16	21
57	787	16	23
58	748	18	25
59	705	20	27

Example 1.3.2 Table 1.9 gives the number of survivors and number of deaths due to two causes. Obtain the chance of decrement due to cause 1 and cause 2 and also the survival probability for all the ages. Compute the probabilities $_2p_{55}^{(\tau)}$, $_{2|}q_{53}^{(1)}$, and $_2q_{56}^{(2)}$.

Solution We compute the decrement probabilities and the survival probability using following R commands. Suppose that Table 1.9 is saved as a file m3.txt on drive D.

```
z <- read.table("D://m3.txt", header=T);
x <- z[, 1];
l <- z[, 2]    #l_x^(τ);
d1 <- z[, 3]   #d_x^(1);
d2 <- z[, 4]   #d_x^(2);
```

Table 1.10 Decrement and survival probabilities

x	$q_x^{(1)}$	$q_x^{(2)}$	$q_x^{(\tau)}$	$p_x^{(\tau)}$
50	0.0100	0.0150	0.0250	0.9750
51	0.0113	0.0164	0.0277	0.9723
52	0.0127	0.0169	0.0295	0.9705
53	0.0141	0.0185	0.0326	0.9674
54	0.0146	0.0202	0.0348	0.9652
55	0.0175	0.0233	0.0407	0.9593
56	0.0194	0.0255	0.0449	0.9551
57	0.0203	0.0292	0.0496	0.9504
58	0.0241	0.0334	0.0575	0.9425
59	0.0284	0.0383	0.0667	0.9333

```
q1 <- d1/1    # q_x^{(1)};
q2 <- d2/1    # q_x^{(2)};
q <- q1+q2    # q_x^{(\tau)};
p <- 1-q   # p_x^{(\tau)};
y <- data.frame(q1, q2, q, p);
y1 <- round(y, 4);
y2 <- data.frame(x, y1);
y2   #Table 1.10;
```

We compute the required probabilities as follows:

$$_2p_{55}^{(\tau)} = p_{55}^{(\tau)} p_{56}^{(\tau)} = (0.9593)(0.9551) = 0.9162,$$

$$_{2|}q_{53}^{(1)} = p_{53}^{(\tau)} p_{54}^{(\tau)} q_{55}^{(1)} = (0.9674)(0.9652)(0.0175) = 0.0163,$$

$$_2q_{56}^{(2)} = q_{56}^{(2)} + p_{56}^{(\tau)} q_{57}^{(2)} = 0.0255 + (0.9551)(0.0292) = 0.0534.$$

The given data may be used to obtain the same probabilities. The answers agree to four decimal places:

$$_2p_{55}^{(\tau)} = \frac{l_{57}^{(\tau)}}{l_{55}^{(\tau)}} = \frac{787}{859} = 0.9162, \qquad _{2|}q_{53}^{(1)} = \frac{d_{55}^{(1)}}{l_{53}^{(\tau)}} = \frac{15}{920} = 0.0163,$$

$$_2q_{56}^{(2)} = \frac{d_{56}^{(2)} + d_{57}^{(2)}}{l_{56}^{(\tau)}} = \frac{21 + 23}{824} = 0.0534.$$

The sets of R commands given in Examples 1.3.1 and 1.3.2 are useful to construct the multiple decrement tables for both the approaches for large data sets, large in the sense of data on more number of ages.

Table 1.11 Double decrement table

x	$q_x^{(1)}$	$q_x^{(2)}$	$q_x^{(\tau)}$	$l_x^{(\tau)}$	$d_x^{(1)}$	$d_x^{(2)}$
40	–	–	0.0075	–	–	12
41	0.002	0.005	–	1800	–	–
42	–	–	–	–	5	–

Example 1.3.3 Calculate $_3q_{40}^{(1)}$ for a double-decrement model, on the basis of information presented in Table 1.11.

Solution By definition, $_3q_{40}^{(1)} = (d_{40}^{(1)} + d_{41}^{(1)} + d_{42}^{(1)})/l_{40}^{(\tau)}$. From the given information,

$$p_{40}^{(\tau)} = 1 - q_{40}^{(\tau)} = 0.9925 = \frac{l_{41}^{(\tau)}}{l_{40}^{(\tau)}} = \frac{1800}{l_{40}^{(\tau)}} \quad \Rightarrow \quad l_{40}^{(\tau)} = 1813.60.$$

Hence,

$$d_{40}^{(\tau)} = l_{40}^{(\tau)} - l_{41}^{(\tau)} = 13.60 = d_{40}^{(1)} + d_{40}^{(2)} = d_{40}^{(1)} + 12 \quad \Rightarrow \quad d_{40}^{(1)} = 1.60,$$

and

$$d_{41}^{(1)} = l_{41}^{(\tau)} q_{41}^{(1)} = (1800)(0.002) = 3.6.$$

Hence we get

$$_3q_{40}^{(1)} = (1.6 + 3.6 + 5)/1813.60 = 0.0056.$$

From the preceding examples we have noted that in building a multiple decrement model, we need to have data on age and number of decrements due to all causes of decrement for the population under study to estimate $q_x^{(j)}$. Large, well-established employer benefit plans may have such data. In some cases, data are available on the number of decrements only due to a specific cause j. We discuss in the next section how to utilize such information to construct a multiple decrement table. On the contrary, given the multiple decrement table, sometimes it is needed to obtain the number of decrements when only a specific cause of decrement is operative and no other causes is operative. The next section also discusses how to obtain such information from multiple decrement table.

1.4 Associated Single Decrement Model

In the multiple decrement model, different modes of decrement apply different stresses which are modeled through the force of decrement for each cause. In the study of relative values of the decremental stresses, the hypothetical elimination of mortality modes leads to marginal structures. In the associated single decrement model all modes, except one, are eliminated. Thus, for each of the causes of decrement in a multiple decrement model, we define a single decrement model that depends only on a particular cause of decrement and then construct the corresponding decrement table. Such a table is known as the associated single decrement table. The associated single decrement table shows the operation of single decrement independent of others. Each table represents a group of lives reduced continuously by only one decrement. This may appear unrealistic, particularly in considering a group subject to, say, withdrawal but not to death. However, the study is useful both in theory and practice. Such associated single decrement tables are also used in survival analysis (Johnson and Johnson, 1980). The decrement rates in associated single decrement model are useful in planning and modeling future financial and actuarial systems where present modes of decrement may be reduced or eliminated at a future date. In general we cannot directly observe decrements due to single cause when all the forces of decrement are active.

In the following we discuss how the associated single decrement table is useful to construct a multiple decrement table, under certain assumptions. The associated single decrement model functions are defined as follows:

$$_t p_x'^{(j)} = \exp\left[-\int_0^t \mu_{x+s}^{(j)}\, ds\right] \quad \text{and} \quad _t q_x'^{(j)} = 1 - _t p_x'^{(j)}.$$

$_t p_x'^{(j)}$ is the probability of survival of (x) to age $x + t$, when a single force j is operative. $_t q_x'^{(j)}$ is the probability of decrement of (x) in $(x, x + t)$ due to cause j only. $_t q_x'^{(j)}$ is called the net probability of decrement in Biostatistics as it is net of other causes of decrement. It is also known as an independent rate of decrement, because cause j does not compete with other causes in determining $_t q_x'^{(j)}$. It is further called the absolute rate of decrement. Sometimes the word rate is used to avoid the word probability. The symbol $_t q_x^{(j)}$ denotes the probability of decrement due to cause j between ages x and $x + t$ when more than one cause is working, and it differs from $_t q_x'^{(j)}$. Note that $_t p_x^{(\tau)} = \exp(-\int_0^t \mu_{x+s}^{(\tau)}\, ds)$. As $t \to \infty$, $_t p_x^{(\tau)} \to 0$. However, it may not be true for $_t p_x'^{(j)}$ for all j. To clarify on this, note that

$$\int_0^\infty \mu_{x+s}^{(\tau)}\, ds = \sum_{j=1}^m \int_0^\infty \mu_{x+s}^{(j)}\, ds.$$

Thus, $\int_0^\infty \mu_{x+s}^{(\tau)}\, ds = \infty \Rightarrow \int_0^\infty \mu_{x+s}^{(j)}\, ds$ is ∞ for at least one j and not necessarily for all j. Thus there may exist j such that $\int_0^\infty \mu_{x+s}^{(j)}\, ds$ is finite, and for that j,

$$_t p_x'^{(j)} = \exp\left(-\int_0^t \mu_{x+s}^{(j)}\, ds\right) \not\to 0 \quad \text{as } t \to \infty.$$

Thus, for some j, $_t p_x'^{(j)}$ may not be a proper survival function, and in the long run there may be a positive number of individuals who would not die due to cause j, which seems reasonable.

We now study the basic relationships between associated rates and multiple decrement model functions. By definition,

$$
{}_t p_x^{(\tau)} = \exp\left\{-\int_0^t \mu_{x+s}^{(\tau)} \, ds\right\} = \exp\left\{-\int_0^t \sum_{j=1}^m \mu_{x+s}^{(j)} \, ds\right\}
$$

$$
= \prod_{j=1}^m \exp\left\{-\int_0^t \mu_{x+s}^{(j)} \, ds\right\} = \prod_{j=1}^m {}_t p_x'^{(j)}.
$$

We know that ${}_t p_x'^{(j)} \in (0, 1)$. Hence, ${}_t p_x^{(\tau)} \leq {}_t p_x'^{(j)}$ for any j. Further,

$$
{}_t p_x'^{(j)} \geq {}_t p_x^{(\tau)} \quad \text{for any } j \quad \Rightarrow \quad {}_t p_x'^{(j)} \mu_{x+t}^{(j)} \geq {}_t p_x^{(\tau)} \mu_{x+t}^{(j)}
$$

Therefore,

$$
\int_0^1 {}_t p_x'^{(j)} \mu_{x+t}^{(j)} \, dt \geq \int_0^1 {}_t p_x^{(\tau)} \mu_{x+t}^{(j)} \, dt = q_x^{(j)}.
$$

It is to be noted that

$$
\int_0^1 {}_t p_x'^{(j)} \mu_{x+t}^{(j)} \, dt = -\int_0^1 \frac{d}{dt} {}_t p_x'^{(j)} 1 \, dt = -\left[{}_t p_x'^{(j)}\right]_0^1 + \int_0^1 {}_t p_x'^{(j)} \frac{d}{dt} 1
$$

$$
= 1 - p_x'^{(j)} = q_x'^{(j)}.
$$

Thus we have proved that $q_x'^{(j)} \geq q_x^{(j)}$. We know that $q_x'^{(j)}$ is the net probability of decrement due to only cause j in one year, while $q_x^{(j)}$ is the crude probability of decrement due to cause j for one year, when some other causes of decrement are operative. Thus, $q_x^{(j)} \leq q_x'^{(j)}$ implies that the probability of death due to jth cause when all are operative is less than the probability of decrement due to cause j when only cause j is operative, which is quite reasonable. The magnitude of other forces of decrement can cause ${}_t p_x'^{(j)}$ to be considerably greater than ${}_t p_x^{(\tau)}$ and leads to the corresponding differences between the absolute rates and the probabilities of decrement. It is to be noted that ${}_t p_x^{(\tau)} \leq {}_t p_x'^{(j)}$ implies ${}_t q_x^{(\tau)} \geq {}_t q_x'^{(j)}$, and hence we get $q_x^{(j)} \leq q_x'^{(j)} \leq {}_t q_x^{(\tau)}$. In the following example we find another upper bound for $q_x'^{(j)}$.

Example 1.4.1 Prove that $q_x'^{(j)} \leq 1 - \exp\{-q_x^{(j)}/(1 - q_x^{(\tau)})\}$.

Solution By definition,

$$
p_x'^{(j)} = \exp\left\{-\int_0^1 \mu_{x+t}^{(j)}\right\}
$$

$$
= \exp\left[\left({}_t p_x^{(\tau)}\right)^{-1}\left\{-\int_0^1 \mu_{x+t}^{(j)} {}_t p_x^{(\tau)}\right\}\right]
$$

Table 1.12 The extract from a triple-decrement table

x	$q_x^{(1)}$	$q_x^{(2)}$	$q_x^{(3)}$	$q_x^{(\tau)}$	$l_x^{(\tau)}$	$q_x'^{(1)}$	$q_x'^{(2)}$	$q_x'^{(3)}$
50	0.001	0.005	0.002	–	10000	–	–	–
51	–	–	–	0.0076	–	–	–	–
52	–	–	–	–	–	0.0023	0.0033	0.0010
53	–	–	–	0.0098	–	–	–	–

$$= \exp\left[-\sum_{r=1}^{\infty} \left({}_tq_x^{(\tau)}\right)^{r-1} q_x^{(j)} \right].$$

We use the fact that ${}_tq_x^{(\tau)}$ is a distribution function; hence, for $0 \le t \le 1$, we have ${}_tq_x^{(\tau)} \le {}_1q_x^{(\tau)} = q_x^{(\tau)}$. Hence,

$$\sum_{r=1}^{\infty} \left({}_tq_x^{(\tau)}\right)^{r-1} \le \sum_{r=1}^{\infty} \left(q_x^{(\tau)}\right)^{r-1} = \left(1 - q_x^{(\tau)}\right)^{-1}.$$

Thus,

$$p_x'^{(j)} \ge \exp\left\{ -q_x^{(j)}/\left(1 - q_x^{(\tau)}\right) \right\}.$$

Hence we have proved that $q_x'^{(j)} \le 1 - \exp\{-q_x^{(j)}/(1 - q_x^{(\tau)})\}$.

The following examples illustrate the basic relationships between associated rates and multiple decrement model functions.

Example 1.4.2 Calculate $l_{53}^{(\tau)}$ given the extract in Table 1.12 from a triple-decrement table.

Solution From the given data we get:

$$p_{50}^{(\tau)} = 1 - (q_{50}^{(1)} + q_{50}^{(2)} + q_{50}^{(3)}) = 0.992, \qquad p_{51}^{(\tau)} = 1 - q_{51}^{(\tau)} = 0.9924,$$

$$p_{52}^{(\tau)} = p_{52}'^{(1)} p_{52}'^{(2)} p_{52}'^{(3)} = (0.9977)(0.9967)(0.9990) = 0.9932.$$

Then

$$\frac{l_{53}^{(\tau)}}{l_{50}^{(\tau)}} = {}_3p_{50}^{(\tau)} = p_{50}^{(\tau)} p_{51}^{(\tau)} p_{52}^{(\tau)} = 0.9778 \quad \Rightarrow \quad l_{53}^{(\tau)} = 9778.$$

Example 1.4.3 For a double-decrement table, it is given that $q_x'^{(2)} = 0.008$, ${}_{1|}q_x^{(1)} = 0.0025$, and $q_{x+1}'^{(1)} = 0.0067$. Calculate $q_x'^{(1)}$.

Solution We have ${}_{1|}q_x^{(1)} = p_x^{(\tau)} q_{x+1}^{(1)}$. Hence, we get $p_x^{(\tau)} = 0.3731$. Now,

$$p_x^{(\tau)} = 0.3731 = \left(1 - q_x'^{(1)}\right)\left(1 - q_x'^{(2)}\right) = \left(1 - q_x'^{(1)}\right)(0.992) \quad \Rightarrow \quad q_x'^{(1)} = 0.6239.$$

Example 1.4.4 Calculate $q_x'^{(2)}$ if

(i) $q_x'^{(2)} = 2q_x'^{(1)}$,
(ii) $q_x'^{(1)} + q_x'^{(2)} = q_x^{(\tau)} + 0.0018$.

Solution For a double-decrement model,

$$q_x^{(\tau)} = 1 - p_x^{(\tau)} = 1 - p_x'^{(1)} p_x'^{(2)} = q_x'^{(1)} + q_x'^{(2)} - q_x'^{(1)} q_x'^{(2)}.$$

From the given information we get

$$q_x'^{(1)} q_x'^{(2)} = 0.0018 \quad \Rightarrow \quad \frac{1}{2} q_x'^{(2)} q_x'^{(2)} = 0.0018 \quad \Rightarrow \quad q_x'^{(2)} = 0.06.$$

Example 1.4.5 From a double-decrement table we have the following data:

(i) $l_{40}^{(\tau)} = 1000$,
(ii) $q_{40}'^{(1)} = 0.002$,
(iii) $q_{40}'^{(2)} = 0.004$,
(iv) $_{1|}q_{40}^{(1)} = 0.005$,
(v) $l_{42}^{(\tau)} = 850$.

Calculate $q_{41}^{(2)}$.

Solution From $q_{40}'^{(1)}$ and $q_{40}'^{(2)}$ we get $p_{40}^{(\tau)} = p_{40}'^{(1)} p_{40}'^{(2)} = 0.9940$. From $l_{40}^{(\tau)}$ and $l_{42}^{(\tau)}$ we get $_2 p_{40}^{(\tau)} = \frac{l_{42}^{(\tau)}}{l_{40}^{(\tau)}} = 0.850$, and then from $_2 p_{40}^{(\tau)} = p_{40}^{(\tau)} p_{41}^{(\tau)}$ we get $p_{41}^{(\tau)} = \frac{_2 p_{40}^{(\tau)}}{p_{40}^{(\tau)}} = 0.8551$ and $q_{41}^{(\tau)} = 0.1449$. From $p_{40}^{(\tau)}$ and $_{1|}q_{40}^{(1)} = p_{40}^{(\tau)} q_{41}^{(1)}$ we get $q_{41}^{(1)} = \frac{_{1|}q_{40}^{(1)}}{p_{40}^{(\tau)}} = 0.0050$. Then $q_{41}^{(2)} = q_{41}^{(\tau)} - q_{41}^{(1)} = 0.1399$.

To summarize, if $\mu_x^{(j)}$ is known for all x, then one can find $_t p_x'^{(j)}$ and $_t q_x'^{(j)}$ for all x and t and can construct the associated single decrement table. In some situations, one may not have information about $\mu_x^{(j)}$ but have information on $q_x^{(j)}$. Thus, it is not possible to obtain $q_x'^{(j)}$ directly. These need to be obtained from $q_x^{(j)}$. We have obtained some inequalities between $q_x'^{(j)}$ and $q_x^{(j)}$. In the following, we discuss how to obtain a relation between $q_x'^{(j)}$ and $q_x^{(j)}$ under some assumptions for fractional ages so that one can be obtained from the other. Commonly used two assumptions are: (i) the constant force of decrement assumption in a unit interval and (ii) the uniform distribution assumption in a unit interval for multiple decrements.

Constant Force of Decrement Assumption in a Unit Age Interval   Under this assumption, $\mu_{x+t}^{(j)} = \mu_x^{(j)}$ and hence $\mu_{x+t}^{(\tau)} = \mu_x^{(\tau)}$ for $0 \leq t < 1$ and x an integer.

Then

$$q_x^{(j)} = \int_0^1 {}_t p_x^{(\tau)} \mu_x^{(j)} \, dt = \frac{\mu_x^{(j)}}{\mu_x^{(\tau)}} \int_0^1 {}_t p_x^{(\tau)} \mu_x^{(\tau)} \, dt = \frac{\mu_x^{(j)}}{\mu_x^{(\tau)}} q_x^{(\tau)}.$$

Further,

$$p_x^{(\tau)} = \exp\left\{ -\int_0^1 \mu_{x+s}^{(\tau)} \, ds \right\} = \exp\left(-\mu_x^{(\tau)}\right) \quad \Rightarrow \quad -\log p_x^{(\tau)} = \mu_x^{(\tau)}.$$

Similarly, $\mu_x^{(j)} = -\log p_x^{\prime(j)}$. So,

$$q_x^{(j)} = \frac{\mu_x^{(j)}}{\mu_x^{(\tau)}} q_x^{(\tau)} = \frac{-\log p_x^{\prime(j)}}{-\log p_x^{(\tau)}} q_x^{(\tau)} = q_x^{(\tau)} \frac{\log p_x^{\prime(j)}}{\log p_x^{(\tau)}}.$$

Thus, if $q_x^{\prime(j)}$, $j = 1, 2, \ldots, m$, are known, $p_x^{\prime(j)} = 1 - q_x^{\prime(j)}$ can be obtained, and then

$$p_x^{(\tau)} = \prod_{j=1}^m p_x^{\prime(j)} \quad \text{and} \quad q_x^{(\tau)} = 1 - p_x^{(\tau)}$$

are obtained. Hence, $q_x^{(j)} = q_x^{(\tau)} \log p_x^{\prime(j)} / \log p_x^{(\tau)}$ can be obtained from the knowledge of $q_x^{\prime(j)}$, $j = 1, \ldots, m$. The identity

$$q_x^{(j)} = \frac{\log p_x^{\prime(j)}}{\log p_x^{(\tau)}} q_x^{(\tau)}$$

can be inverted to obtain $q_x^{\prime(j)}$ from $q_x^{(j)}$. The above identity can be rewritten as

$$\log p_x^{\prime(j)} = \frac{q_x^{(j)}}{q_x^{(\tau)}} \log p_x^{(\tau)} = \log\left[\left(p_x^{(\tau)}\right)^{\frac{q_x^{(j)}}{q_x^{(\tau)}}}\right].$$

Therefore,

$$p_x^{\prime(j)} = \left(p_x^{(\tau)}\right)^{\frac{q_x^{(j)}}{q_x^{(\tau)}}} \quad \text{and} \quad q_x^{\prime(j)} = 1 - \left(1 - q_x^{(\tau)}\right)^{\frac{q_x^{(j)}}{q_x^{(\tau)}}}.$$

Thus absolute rates of decrement can be obtained from a given set of probabilities of decrement and vice versa. We now study the second assumption to obtain such relations and the uniform distribution assumption for multiple decrements in a unit age interval.

Uniform Distribution Assumption in a Unit Age Interval   Under this assumption, each of the decrements in a multiple decrement context satisfy a uniform distribution assumption in each year of age. Hence, ${}_t q_x^{(j)} = t q_x^{(j)}$, $j = 1, 2, \ldots, m$,

$0 \le t \le 1$, and x is an integer. As a consequence,

$$_t q_x^{(\tau)} = \sum_{j=1}^{m} {_t q^{(j)x}} = t \sum_{j=1}^{m} q_x^{(j)} = t q^{(\tau)x}.$$

Further,

$$\mu_{x+t}^{(j)} = \frac{1}{_t p_x^{(\tau)}} \frac{d}{dt} {_t q_x^{(j)}} = \frac{1}{_t p_x^{(\tau)}} \frac{d}{dt} t q_x^{(j)} = \frac{1}{_t p_x^{(\tau)}} q_x^{(j)} = \frac{q_x^{(j)}}{(1 - t q_x^{(\tau)})}.$$

Hence,

$$q_x'^{(j)} = 1 - p_x'^{(j)} = 1 - \exp\left\{-\int_0^1 \mu_{x+t}^{(j)} \, dt\right\}$$

$$= 1 - \exp\left\{-\int_0^1 \frac{q_x^{(j)}}{(1 - t q_x^{(\tau)})} \, dt\right\}$$

$$= 1 - \exp\left\{\frac{q_x^{(j)}}{q_x^{(\tau)}} \left[\log(1 - t q_x^{(\tau)})\right]_0^1\right\}$$

$$= 1 - \exp\left\{\frac{q_x^{(j)}}{q_x^{(\tau)}} \log(1 - q_x^{(\tau)})\right\}$$

$$= 1 - (1 - q_x^{(\tau)})^{q_x^{(j)}/q_x^{(\tau)}},$$

which is exactly the same relation as under the constant force of decrement assumption.

Thus, once associated single decrement probabilities are known, the results derived above can be used to construct the multiple decrement table. The availability of a set of $p_x'^{(j)}$ for $j = 1, \ldots, m$ and for all values of x will permit the computations of $p_x^{(\tau)}$ and of $q_x^{(\tau)}$. The next step is to break $q_x^{(\tau)}$ into components $q_x^{(j)}$ for $j = 1, \ldots, m$. If either the constant force or the uniform distribution of decrement assumption in a unit age interval is adopted in the model, the $q_x^{(j)}$ values can be obtained using the results derived above. The following examples illustrate the method.

Example 1.4.6 Given that decrement may be due to death, 1, disability, 2, or retirement, 3, use constant force of decrement assumption for each unit age interval to construct a multiple decrement table based on the absolute rates (see Table 1.13).

Solution From the given data we find

$$p_x^{(\tau)} = (1 - q_x'^{(1)})(1 - q_x'^{(2)})(1 - q_x'^{(3)}), \qquad q_x^{(\tau)} = 1 - p_x^{(\tau)}$$

$$\text{and} \quad q_x^{(j)} = q_x^{(\tau)} \frac{\log p_x'^{(j)}}{\log p_x^{(\tau)}}, \qquad j = 1, 2, 3.$$

Table 1.13 Associated single decrement probabilities

Age x	$q_x'^{(1)}$	$q_x'^{(2)}$	$q_x'^{(3)}$
55	0.0210	0.029	0.20
56	0.0215	0.030	0.10
57	0.0220	0.033	0.13
58	0.0230	0.034	0.12
59	0.0260	0.038	0.14

Table 1.14 Multiple decrement table

Age x	$p_x^{(\tau)}$	$q_x^{(1)}$	$q_x^{(2)}$	$q_x^{(3)}$
55	0.7605	0.0186	0.0257	0.1952
56	0.8542	0.0201	0.0282	0.0975
57	0.8228	0.0202	0.0305	0.1265
58	0.8305	0.0212	0.0316	0.1167
59	0.8058	0.0237	0.0348	0.1357

The following R commands give us the required probabilities. Suppose that the given data are stored in a file m5.txt on D drive.

```
z <- read.table("D://m5.txt", header=T);
x <- z[, 1];    q1 <- z[, 2];    q2 <- z[, 3];
q3 <- z[, 4];     p1 <- 1-q1;    p2 <- 1-q2;
p3 <- 1-q3;    p <- p1*p2*p3;    q <- 1-p;
e <- exp(1)
a1 <- q*log(p1, base=e)/log(p, base=e);
a2 <- q*log(p2, base=e)/log(p, base=e);
a3 <- q*log(p3, base=e)/log(p, base=e);
y <- round(data.frame(p, a1, a2, a3), 4);
y1 <- data.frame(x, y);    y1   #Table 1.14;
```

The results are summarized in Table 1.14.

Example 1.4.7 Three forces of mortality for (x) are given below:

$$\mu_{x+s}^{(1)} = 0.003 + 0.0024(x + s - 40)^2, \qquad \mu_{x+s}^{(2)} = 0.003 + 0.0007(x + s - 40)^{2.5},$$

$$\mu_{x+s}^{(3)} = 0.003 + 0.00004(x + s - 40)^3, \quad s \geq 0, \ x \geq 40.$$

Obtain the associated single decrement table for ages 50 to 60. Under the assumption of uniformity for fractional ages, find the corresponding multiple decrement table.

Solution We have, by definition,

$$_tp_x'^{(j)} = \exp\left[-\int_0^t \mu_{x+s}^{(j)}\,ds\right] \quad \text{and} \quad _tp_x^{(\tau)} = \prod_{j=1}^m {_tp_x'^{(j)}}.$$

Under the assumption of uniformity for fractional ages, we have

$$q_x^{(j)} = q_x^{(\tau)}\frac{\log p_x'^{(j)}}{\log p_x^{(\tau)}}, \quad j = 1, 2, 3.$$

For the given forces of mortality we find:

$$_tp_x'^{(1)} = \exp\{-[0.003t + 0.0008((x+t-40)^3 - (x-40)^3)]\},$$

$$_tp_x'^{(2)} = \exp\{-[0.003t + 0.0002((x+t-40)^{3.5} - (x-40)^{3.5})]\}, \quad \text{and}$$

$$_tp_x'^{(3)} = \exp\{-[0.003t + 0.00001((x+t-40)^4 - (x-40)^4)]\}.$$

We use these formulae to find the first associated single decrement table and then the corresponding multiple decrement table:

```
e <- exp(1);
x <- seq(50, 60, 1);
p1 <- e^(-(0.003+0.0008*((x+1-40)^3-(x-40)^3)))
                                          #a vector of p_x'^(1);
p2 <- e^(-(0.003+0.0002*((x+1-40)^3.5-(x-40)^(3.5))))
        #a vector of p_x'^(2);
p3 <- e^(-(0.003+0.00001*((x+1-40)^4-(x-40)^4)))
        #a vector of p_x'^(3);
q1 <- 1-p1   #a vector of q_x'^(1);
q2 <- 1-p2   #a vector of q_x'^(2);
q3 <- 1-p3   #a vector of q_x^/(3);
p <- p1*p2*p3   #a vector of p_x^(τ);
q <- 1-p   #a vector of q_x^(τ);
a1 <- q*log(p1, base=e)/log(p, base=e) #a vector of q_x^(1);
a2 <- q*log(p2, base=e)/log(p, base=e) #a vector of q_x^(2);
a3 <- q*log(p3, base=e)/log(p, base=e) #a vector of q_x^(3);
y <- round(data.frame(q1, q2, q3, p, a1, a2, a3), 4);
y1 <- data.frame(x, y);
y1   #Table 1.15;
```

From Table 1.15 we note that after age 52, $q_x'^{(2)}$ is higher than $q_x'^{(1)}$ and $q_x'^{(3)}$. It is in view of the fact that the force of mortality corresponding to cause 2 is higher after this age. This is clear from the graph of three forces of decrement shown in Fig. 1.5. Further, $q_x'^{(j)}$ is higher than $q_x^{(j)}$ for $j = 1, 2, 3$ and $x = 50, 51, \ldots, 60$.

Table 1.15 Multiple decrement table from associated single decrement model

Age x	$q_x'^{(1)}$	$q_x'^{(2)}$	$q_x'^{(3)}$	$p_x^{(\tau)}$	$q_x^{(1)}$	$q_x^{(2)}$	$q_x^{(3)}$
50	0.2349	0.2239	0.0482	0.5652	0.2041	0.1931	0.0377
51	0.2743	0.2719	0.0619	0.4957	0.2304	0.2280	0.0460
52	0.3149	0.3230	0.0780	0.4276	0.2548	0.2628	0.0547
53	0.3563	0.3763	0.0966	0.3627	0.2769	0.2967	0.0638
54	0.3982	0.4310	0.1176	0.3022	0.2961	0.3288	0.0729
55	0.4400	0.4860	0.1411	0.2472	0.3123	0.3585	0.0819
56	0.4814	0.5405	0.1671	0.1985	0.3255	0.3854	0.0906
57	0.5220	0.5935	0.1955	0.1563	0.3356	0.4092	0.0989
58	0.5616	0.6444	0.2262	0.1206	0.3429	0.4299	0.1066
59	0.5998	0.6924	0.2590	0.0912	0.3476	0.4474	0.1138
60	0.6364	0.7370	0.2938	0.0675	0.3501	0.4620	0.1203

Fig. 1.5 Forces of decrement

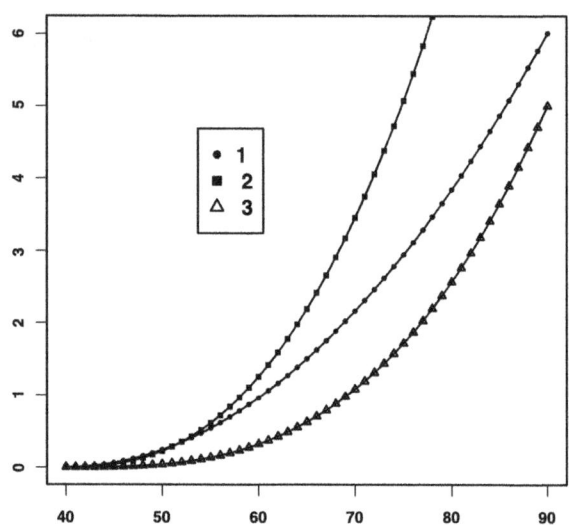

Example 1.4.8 For a triple-decrement model, the values of $\mu_x^{(j)}$ for $j = 1, 2, 3$ are $0.03, 0.04$, and 0.05, respectively. Calculate

(i) $q_x^{(j)}$ and $q_x'^{(j)}$ for $j = 1, 2, 3$.
(ii) the upper bound on $q_x'^{(j)}$ for $j = 1, 2, 3$, as derived in Example 1.4.1.
(iii) $q_x'^{(j)}$ for $j = 1, 2, 3$ under the assumption of uniformity of deaths in a unit age interval.
(iv) Compare the values in (iii) with the exact values and the upper bound calculated in (i) and (ii).

Table 1.16 Values of associated single decrement rates

j	Exact value	Approximate value	Upper bound
1	0.02827	0.02958	0.03142
2	0.03769	0.03924	0.04332
3	0.04712	0.04881	0.05183

Solution (i) We have $\mu_{x+t}^{(\tau)} = \sum_j \mu_{x+t}^{(j)} = 0.12$. Hence,

$$_t p_x^{(\tau)} = \exp\left\{-\int_0^t \mu_{x+s}^{(\tau)} \, ds\right\} = \exp(-0.12t).$$

By definition,

$$q_x^{(1)} = \int_0^1 {}_t p_x^{(\tau)} \mu_{x+t}^{(1)} \, dt = \int_0^1 e^{-0.12t}(0.03) \, dt = \frac{1}{4}\left(1 - e^{-0.12}\right) = 0.02827.$$

On similar lines we get, $q_x^{(2)} = \frac{1}{3}(1 - e^{-0.12}) = 0.03769$ and $q_x^{(3)} = \frac{5}{12}(1 - e^{-0.12}) = 0.04712$. Further,

$$q_x'^{(1)} = 1 - \exp\left\{\int_0^1 \mu_{x+t}^{(1)} \, dt\right\} = 1 - \exp(-0.03) = 0.02955.$$

Similarly we get $q_x'^{(2)} = 0.03921$ and $q_x'^{(3)} = 0.04877$. It is to be noted that $q_x'^{(j)} \geq q_x^{(j)}$ for $j = 1, 2, 3$.

(ii) The upper bound of the $q_x'^{(j)}$ derived in Example 1.4.1 is given by

$$q_x'^{(j)} \leq 1 - \exp\left\{-q_x^{(j)}/\left(1 - q_x^{(\tau)}\right)\right\}.$$

So we get

$$q_x'^{(1)} \leq 0.03142, \qquad q_x'^{(2)} \leq 0.04332, \qquad q_x'^{(3)} \leq 0.05183.$$

(iii) Under the assumption of uniformity in unit age interval, we have

$$q_x'^{(j)} = 1 - \left(1 - q_x^{(\tau)}\right)^{\frac{q_x^{(j)}}{q_x^{(\tau)}}}.$$

Using this formula, we get

$$q_x'^{(1)} = 0.02958, \qquad q_x'^{(2)} = 0.03924, \qquad q_x'^{(3)} = 0.04881.$$

(iv) Table 1.16 presents the upper bounds and the exact and approximate values of $q_x'^{(j)}$ under the assumption of uniformity.

Table 1.17 The extract from a double-decrement table

x	$l_x^{'(\tau)}$	$d_x^{(1)}$	$d_x^{(2)}$
35	100	–	–
36	–	2	12
37	–	1	2

It is to be noted that there is a close agreement among the upper bound, the exact and approximate values of $q_x^{'(j)}$ under the assumption of uniformity. Such observation leads to the application of uniformity approximation for the associated single decrement model.

Example 1.4.9 For a double-decrement model, it is given that $p_x^{'(2)} = [p_x^{'(1)}]^2 = e^{-0.2}$ and each decrement has a constant force of decrement over each year of age. Calculate $q_x^{(1)}$ and $q_x^{(2)}$.

Solution From the given information we have

$$e^{-0.2} = p_x^{'(2)} = \exp\left\{-\int_0^1 \mu_{x+s}^{(2)}\, ds\right\} = \exp\{-\mu^{(2)}\}.$$

Hence, $\mu^{(2)} = 0.2$. Similarly, $p_x^{'(1)} = e^{-0.1}$ gives $\mu^{(1)} = 0.1$. Consequently, $\mu_x^{(\tau)} = 0.3$ and $p_x^{(\tau)} = e^{-0.3}$ giving $q_x^{(\tau)} = 1 - e^{-0.3}$. Further, under the assumption of constant force of mortality,

$$q_x^{(1)} = \frac{\mu_x^{(1)}}{\mu_x^{(\tau)}} q_x^{(\tau)} = \frac{1}{3}(1 - e^{-0.3}) = 0.08639$$

and

$$q_x^{(2)} = \frac{\mu_x^{(2)}}{\mu_x^{(\tau)}} q_x^{(\tau)} = \frac{2}{3}(1 - e^{-0.3}) = 0.1728.$$

Example 1.4.10 For a double-decrement model, it is given that the force of decrement is constant for each decrement over the year of age 35, 36, and 37, $q_{35}^{'(1)} = 0.01$, $q_{35}^{'(2)} = 0.06$, and further information is given in Table 1.17. Find $q_{37}^{'(1)}$.

Solution From the given information, $p_{35}^{(\tau)} = (1 - q_{35}^{'(1)})(1 - q_{35}^{'(2)}) = 0.9306$. Hence,

$$l_{36}^{(\tau)} = 100(0.9306) = 93.06 \quad \text{and} \quad l_{37}^{(\tau)} = l_{36}^{(\tau)} - d_{36}^{(1)} - d_{36}^{(2)} = 93.06 - 2 - 12 = 79.06.$$

Now,

$$q_{37}^{(1)} = d_{37}^{(1)}/l_{37}^{(\tau)} = 0.01265 \quad \text{and} \quad q_{37}^{(\tau)} = d_{37}^{(\tau)}/l_{37}^{(\tau)} = 3/79.06 = 0.03795.$$

Table 1.18 The extract from a double-decrement table

x	$l_x^{(\tau)}$	$d_x^{(1)}$	$d_x^{(2)}$
45	1000	36	42
46	–	–	68
47	820	–	–

Hence,

$$q_{37}'^{(1)} = 1 - \left[1 - q_{37}^{(\tau)}\right]^{q_{37}^{(1)}/q_{37}^{(\tau)}} = 0.01266.$$

Example 1.4.11 For a triple-decrement model, it is given that

(i) $q_{50}^{(1)} = q_{50}^{(3)}$,
(ii) $q_{50}^{(2)} = 2q_{50}^{(1)}$, and
(iii) $\mu_{50+t}^{(1)} = \log 8, 0 \le t < 1$.

Assume a constant force of decrement for each decrement over each year of age. Calculate $q_{50}'^{(2)}$.

Solution Under the assumption of constant force of decrement,

$$\frac{q_x^{(j)}}{q_x^{(i)}} = \frac{\mu_x^{(j)}}{\mu_x^{(i)}} \quad \Rightarrow \quad 2 = \frac{q_{50}^{(2)}}{q_{50}^{(1)}} = \frac{\mu_{50}^{(2)}}{\mu_{50}^{(1)}}$$

$$\Rightarrow \quad \mu_{50}^{(2)} = 2\mu_{50}^{(1)} = 2\log 8 = \log 64$$

$$\Rightarrow \quad p_{50}'^{(2)} = e^{-\mu_{50}^{(2)}} = e^{-\log 64} = \frac{1}{64}$$

$$\Rightarrow \quad q_{50}'^{(2)} = 0.9844.$$

Example 1.4.12 For a double-decrement model, the data are given in Table 1.18. Assume that each decrement is uniformly distributed over each year of age. Calculate the absolute rate of decrement due to cause 1 for age 46.

Solution We have to find $q_{46}'^{(1)} = 1 - [p_{46}^{(\tau)}]^{q_{46}^{(1)}/q_{46}^{(\tau)}}$. From the given data we see that $l_{46}^{(\tau)} = l_{45}^{(\tau)} - d_{45}^{(1)} - d_{45}^{(2)} = 922$. Further,

$$820 = l_{47}^{(\tau)} = l_{46}^{(\tau)} - d_{46}^{(1)} - d_{46}^{(2)} = 922 - d_{46}^{(1)} - 68 \quad \Rightarrow \quad d_{46}^{(1)} = 34.$$

Then,

$$\frac{q_{46}^{(1)}}{q_{46}^{(\tau)}} = \frac{d_{46}^{(1)}}{d_{46}^{(\tau)}} = \frac{34}{34 + 68} = \frac{1}{3} \quad \text{and} \quad p_{46}^{(\tau)} = \frac{820}{922} = 0.8894 \quad \Rightarrow \quad q_{46}'^{(1)} = 0.03832.$$

Example 1.4.13 For a triple-decrement model, it is assumed that each decrement is uniformly distributed over each year of age in its associated single decrement table. Prove that

$$q_x^{(1)} = q_x'^{(1)}\left[1 - \frac{1}{2}\left(q_x'^{(2)} + q_x'^{(3)}\right) + \frac{1}{3}q_x'^{(2)}q_x'^{(3)}\right].$$

Solution Under the assumption of uniformity in the associated single decrement table, $_tq_x'^{(j)} = t q_x'^{(j)}$ and $_t p_x'^{(1)} \mu_{x+t}^{(1)} = q_x'^{(1)}$. By definition,

$$q_x^{(1)} = \int_0^1 {}_t p_x^{(\tau)} \mu_{x+t}^{(1)}\, dt = \int_0^1 {}_t p_x'^{(1)} \, {}_t p_x'^{(2)} \, {}_t p_x'^{(3)} \mu_{x+t}^{(1)}\, dt$$

$$= q_x'^{(1)} \int_0^1 \left(1 - t q_x'^{(2)}\right)\left(1 - t q_x'^{(3)}\right) dt$$

$$= q_x'^{(1)}\left[1 - \frac{1}{2}\left(q_x'^{(2)} + q_x'^{(3)}\right) + \frac{1}{3}q_x'^{(2)}q_x'^{(3)}\right].$$

Example 1.4.14 For a triple-decrement model, it is assumed that each decrement is uniformly distributed over each year of age in its associated single decrement table. Further, $q_x'^{(1)} = 0.01$, $q_x'^{(2)} = 0.04$, and $q_x'^{(3)} = 0.0625$. Calculate $q_x^{(1)}$.

Solution From Example 1.4.13 we have, under the assumption of uniformity in the associated single decrement table,

$$q_x^{(1)} = q_x'^{(1)}\left[1 - \frac{1}{2}\left(q_x'^{(2)} + q_x'^{(3)}\right) + \frac{1}{3}q_x'^{(2)}q_x'^{(3)}\right] = 0.009496.$$

Example 1.4.15 The following information is given for a double-decrement table.

(i) $\mu_{x+.5}^{(1)} = 0.01$,

(ii) $q_x^{(2)} = 0.001$, and

(iii) each decrement is uniformly distributed over each year of age in its associated single decrement table.

Calculate $q_x^{(1)}$.

Solution Under the given assumption, proceeding as in Example 1.4.13, we get $q_x^{(1)} = q_x'^{(1)}[1 - \frac{1}{2}q_x'^{(2)}]$. To calculate the quantities involved in this expression, under the given assumption, we also have

$$\mu_{x+t}^{(j)} = \frac{q_x'^{(j)}}{{}_t p_x'^{(j)}} = \frac{q_x'^{(j)}}{1 - t \cdot q_x'^{(j)}} \quad \Rightarrow \quad 0.01 = \mu_{x+0.5}^{(1)} = \frac{q_x'^{(1)}}{1 - (0.5)q_x'^{(1)}}$$

$$\Rightarrow \quad q_x'^{(1)} = 0.0099.$$

Also,

$$0.001 = q_x^{(2)} = q_x'^{(2)}\left[1 - \frac{1}{2}q_x'^{(1)}\right] \quad \Rightarrow \quad q_x'^{(2)} = 0.0011.$$

Hence,

$$q_x^{(1)} = q_x'^{(1)}\left[1 - \frac{1}{2}q_x'^{(2)}\right] = (0.0099)\left[1 - \frac{1}{2}(0.0011)\right] = 0.009894.$$

There is one more approach to find $q_x'^{(j)}$ from $q_x^{(j)}$ and vice versa. It consists of defining central rates of multiple decrement. We discuss it below.

Central Rates of Multiple Decrement   The central rate of mortality or the central death rate at age x in a single decrement table, denoted by m_x, is defined as

$$m_x = \frac{\int_0^1 {}_t p_x \mu_{x+t}\, dt}{\int_0^1 {}_t p_x\, dt} = \frac{\int_0^1 l_{x+t}\mu_{x+t}\, dt}{\int_0^1 l_{x+t}\, dt} = \frac{l_x q_x}{\int_0^1 l_{x+t}\, dt} = \frac{d_x}{\int_0^1 l_{x+t}\, dt} = \frac{d_x}{L_x}.$$

Thus, m_x is a weighted average of the force of mortality between ages x and $x + 1$, and this justifies the terminology central rate. Such central rates are defined in a multiple decrement context as follows. The central rate of decrement from all causes is defined by

$$m_x^{(\tau)} = \frac{\int_0^1 {}_t p_x^{(\tau)}\, \mu_{x+t}^{(\tau)}\, dt}{\int_0^1 {}_t p_x^{(\tau)}\, dt} = \frac{l_x^{(\tau)} q_x^{(\tau)}}{\int_0^1 l_{x+t}^{(\tau)}\, dt} = \frac{d_x^{(\tau)}}{\int_0^1 l_{x+t}^{(\tau)}\, dt}.$$

It is a weighted average of $\mu_{x+t}^{(\tau)}$, $0 \le t < 1$. Similarly, the central rate of decrement from cause j is

$$m_x^{(j)} = \frac{\int_0^1 {}_t p_x^{(\tau)} \mu_{x+t}^{(j)}\, dt}{\int_0^1 {}_t p_x^{(\tau)}} = \frac{l_x^{(\tau)} q_x^{(j)}}{\int_0^1 l_{x+t}^{(\tau)}\, dt} = \frac{d_x^{(j)}}{\int_0^1 l_{x+t}^{(\tau)}\, dt}.$$

It is a weighted average of $\mu_{x+t}^{(j)}$. The corresponding central rate for the associated single decrement table is given by

$$m_x'^{(j)} = \int_0^1 {}_t p_x'^{(j)} \mu_{x+t}^{(j)}\, dt \bigg/ \int_0^1 {}_t p_x'^{(j)}\, dt.$$

This is again a weighted average of $\mu_{x+t}^{(j)}$ over the same age range, with weights ${}_t p_x'^{(j)}$ and not ${}_t p_x^{(\tau)}$. From the above definitions of central rates it is easy to see that if the force $\mu_{x+t}^{(j)}$ of decrement is constant for $0 \le t < 1$, we have

$$m_x^{(j)} = m_x'^{(j)} = \mu_x^{(j)}.$$

The following examples illustrate the application of central rate of decrement to find multiple decrement model functions.

Example 1.4.16 In a triple-decrement model, the causes of decrement are death, withdrawal, and disability. It is given that each decrement is uniformly distributed over each year of age. Further, $l_x^{(\tau)} = 10000$, $l_{x+1}^{(\tau)} = 9185$, the central rate of death for all ages equals 0.01, and the central rate of withdrawal for all ages equals 0.05. Calculate the probability of decrement by disability at age x.

Solution Under the assumption of uniformity, $l_{x+s}^{(\tau)} = l_x^{(\tau)} - s(l_x^{(\tau)} - l_{x+1}^{(\tau)})$. Hence, $\int_0^1 l_{x+s}^{(\tau)} ds = \frac{l_x^{(\tau)} + l_{x+1}^{(\tau)}}{2}$. By the definition of central rate of death at age x,

$$0.01 = m_x^{(d)} = \frac{d_x^{(d)}}{\int_0^1 l_{x+s}^{(\tau)} ds} = \frac{2d_x^{(d)}}{l_x^{(\tau)} + l_{x+1}^{(\tau)}} \quad \text{hence,} \quad d_x^{(d)} = 95.925.$$

Similarly, $d_x^{(w)} = 479.625$ and $d_x^{(dis)} = 10000 - 9185 - 95.925 - 479.625 = 239.45$. Hence,

$$q_x^{(dis)} = \frac{239.45}{10000} = 0.0239.$$

Example 1.4.17 For a double-decrement model, it is given that each decrement has a constant force of mortality over each year of age. Further, $m_x^{(1)} = 0.01$ and $m_x^{(2)} = 0.02$. Calculate $q_x^{(1)}$.

Solution Under the constant force of mortality assumption, $m_x^{(j)} = m_x'^{(j)} = \mu_x^{(j)}$. Hence,

$$q_x^{(1)} = \int_0^1 {}_t p_x^{(\tau)} \mu_{x+t}^{(1)} dt = \int_0^1 e^{-(\mu^{(1)} + \mu^{(2)})t} \mu^{(1)} dt$$

$$= \int_0^1 e^{-0.03t} (0.01) dt = 0.00985.$$

In the following example, we establish relations between the central rate of decrements and the decrement probabilities, under certain assumptions. These relations are useful to construct a multiple decrement table using the central rate bridge.

Example 1.4.18 Show that, under the assumption of a uniform distribution of decrements in a multiple decrement model,

(a) $m_x^{(\tau)} = \dfrac{q_x^{(\tau)}}{1 - (1/2)q_x^{(\tau)}}$,

(b) $m_x^{(j)} = \dfrac{q_x^{(j)}}{1 - (1/2)q_x^{(\tau)}}$, and conversely,

(c) $q_x^{(\tau)} = \dfrac{m_x^{(\tau)}}{1+(1/2)m_x^{(\tau)}}$,

(d) $q_x^{(j)} = \dfrac{m_x^{(j)}}{1+(1/2)m_x^{(\tau)}}$.

Show that, under the assumption of a uniform distribution of decrements in the associated single decrement model,

(e) $m_x^{\prime(j)} = \dfrac{q_x^{\prime(j)}}{1-(1/2)q_x^{\prime(j)}}$, and hence,

(f) $q_x^{\prime(j)} = \dfrac{m_x^{\prime(j)}}{1+(1/2)m_x^{\prime(j)}}$.

Solution Under the assumption of uniform distribution of decrements in a multiple decrement model,

$$_tp_x^{(\tau)} = 1 - tq_x^{(\tau)}, \qquad \mu_{x+t\,t}^{(\tau)}p_x^{(\tau)} = q_x^{(\tau)} \quad \text{and} \quad \mu_{x+t\,t}^{(j)}p_x^{(\tau)} = q_x^{(j)}.$$

Hence,

$$\int_0^1 {}_tp_x^{(\tau)}\,dt = \int_0^1 \left(1 - tq_x^{(\tau)}\right)dt = 1 - \frac{1}{2}q_x^{(\tau)}.$$

Therefore,

$$m_x^{(\tau)} = \int_0^1 {}_tp_x^{(\tau)}\mu_{x+t}^{(\tau)}\,dt \Big/ \int_0^1 {}_tp_x^{(\tau)}\,dt = q_x^{(\tau)} \Big/ \left(1 - \frac{1}{2}q_x^{(\tau)}\right),$$

$$m_x^{(j)} = \int_0^1 {}_tp_x^{(\tau)}\mu_{x+t}^{(j)}\,dt \Big/ \int_0^1 {}_tp_x^{(\tau)}\,dt = q_x^{(j)} \Big/ \left(1 - \frac{1}{2}q_x^{(\tau)}\right).$$

Thus (a) and (b) are proved. Now,

$$m_x^{(\tau)} = q_x^{(\tau)} \Big/ \left(1 - \frac{1}{2}q_x^{(\tau)}\right) \quad \Rightarrow \quad \left(1 - \frac{1}{2}q_x^{(\tau)}\right)m_x^{(\tau)} - q_x^{(\tau)} = 0$$

$$\Rightarrow \quad q_x^{(\tau)} = m_x^{(\tau)} \Big/ \left(1 + \frac{1}{2}m_x^{(\tau)}\right).$$

Similarly, $m_x^{(j)} = q_x^{(j)} / (1 - \frac{1}{2}q_x^{(\tau)})$. Hence,

$$q_x^{(j)} = m_x^{(j)}\left(1 - \frac{1}{2}q_x^{(\tau)}\right) = m_x^{(j)}\left(1 - \frac{\frac{1}{2}m_x^{(\tau)}}{1 + \frac{1}{2}m_x^{(\tau)}}\right) = m_x^{(j)} \Big/ \left(1 + \frac{1}{2}m_x^{(\tau)}\right).$$

To prove (e), note that by definition, $m_x^{\prime(j)} = \int_0^1 {}_tp_x^{\prime(j)}\mu_{x+t}^{(j)} \big/ \int_0^1 {}_tp_x^{\prime(j)}\,dt$. Under the assumption of uniformity in the associated single decrement model, $_tq_x^{\prime(j)} = tq_x^{\prime(j)}$.

Table 1.19 The absolute rates

Age x	$q_x'^{(1)}$	$q_x'^{(2)}$	$q_x'^{(3)}$
55	0.0210	0.029	0.20
56	0.0215	0.030	0.10
57	0.0220	0.033	0.13
58	0.0230	0.034	0.12
59	0.0260	0.038	0.14

Hence, $_t p_x'^{(j)} = 1 - {}_t q_x'^{(j)} = 1 - t q_x'^{(j)}$ and $\mu_{x+t}^{(j)} {}_t p_x'^{(j)} = q_x'^{(j)}$. Therefore,

$$m_x'^{(j)} = q_x'^{(j)} / \int_0^1 {}_t p_x'^{(j)} \, dt = q_x'^{(j)} / \left(1 - \frac{1}{2} q_x'^{(j)} \right).$$

From (e) we get $q_x'^{(j)} = m_x'^{(j)} / (1 + \frac{1}{2} m_x'^{(j)})$.

Example 1.4.19 Show that, under the assumption of constant force of decrements in a multiple decrement model, $q_x^{(j)} = m_x^{(j)} (1 - e^{-m_x^{(\tau)}}) / m_x^{(\tau)}$.

Solution Under the assumption of constant force of decrements in a multiple decrement model, $\mu_{x+t}^{(j)} = \mu_x^{(j)}$ for $j = 1, 2, \ldots, m$. As a consequence, $\mu_{x+t}^{(\tau)} = \mu_x^{(\tau)}$. Now,

$$q_x^{(j)} = \int_0^1 {}_t p_x^{(\tau)} \mu_{x+t}^{(j)} = \mu_x^{(j)} \int_0^1 {}_t p_x^{(\tau)}.$$

Further, $_t p_x^{(\tau)} = \exp\{-\int_0^t \mu_{x+s}^{(\tau)} \, ds\} = \exp\{-t \mu_x^{(\tau)}\}$. Hence, $\int_0^1 {}_t p_x^{(\tau)} = (1 - \exp(-\mu_x^{(\tau)})) / \mu_x^{(\tau)}$. But, under the assumption of constant force of decrements, $\mu_x^{(\tau)} = m_x^{(\tau)}$. Hence we get

$$q_x^{(j)} = \mu_x^{(j)} \int_0^1 {}_t p_x^{(\tau)} = m_x^{(j)} (1 - e^{-m_x^{(\tau)}}) / m_x^{(\tau)}.$$

We use these relations in the following example to construct a multiple decrement table.

Example 1.4.20 Given that decrement may be due to death, 1, disability, 2, or retirement, 3, use the constant force of decrement assumption to construct a multiple decrement table based on the absolute rates (Table 1.19). Use the central rate bridge. State the underlying assumptions.

Solution We are given that the force of mortality is constant for all the causes. We further assume the uniformity in associated single decrement model. Then we have

$$m_x'^{(j)} = q_x'^{(j)} / \left(1 - \frac{1}{2} q_x'^{(j)}\right).$$

Further, assume that $m_x'^{(j)} = m_x^{(j)}$, which is valid if the force of mortality remains constant in a unit age interval. Again under the same assumption,

$$q_x^{(j)} = m_x^{(j)}\left(1 - e^{-m_x^{(\tau)}}\right) / m_x^{(\tau)}.$$

Now, to find $m_x^{(\tau)}$, note that

$$m_x^{(\tau)} = \int_0^1 {}_t p_x^{(\tau)} \mu_{x+t}^{(\tau)} \, dt / \int_0^1 {}_t p_x^{(\tau)} \, dt = \int_0^1 {}_t p_x^{(\tau)} \sum_{j=1}^m \mu_{x+t}^{(j)} \, dt / \int_0^1 {}_t p_x^{(\tau)} \, dt$$

$$= \sum_{j=1}^m \int_0^1 {}_t p_x^{(\tau)} \mu_{x+t}^{(j)} \, dt / \int_0^1 {}_t p_x^{(\tau)} \, dt = \sum_{j=1}^m m_x^{(j)}.$$

Thus, from $q_x'^{(j)}$ we get $m_x'^{(j)} = m_x^{(j)}$; then we find $m_x^{(\tau)}$ and hence $q_x^{(j)}$ by using the result proved in Example 1.4.19. Using these steps, the following R commands produce the required results. Suppose that the given data are stored in m5.txt on drive D. These data are the same as in Example 1.4.6.

```
z <- read.table("D://m5.txt", header=T);
x <- z[, 1];
q1 <- z[, 2]   #vector of q'(1)_x;
q2 <- z[, 3]   #vector of q'(2)_x;
q3 <- z[, 4]   #vector of q'(3)_x;
m1 <- q1/(1-0.5*q1)   #vector of m'(1)_x = m(1)_x;
m2 <- q2/(1-0.5*q2)   #vector of m'(2)_x = m(2)_x;
m3 <- q3/(1-0.5*q3)   #vector of m'(3)_x = m(3)_x;
m <- m1+m2+m3   #vector of m(τ)_x;
e <- exp(1);
a <- e^(-m);
a1 <- m1*(1-a)/m   #vector of q(1)_x;
a2 <- m2*(1-a)/m   #vector of q(2)_x;
a3 <- m3*(1-a)/m   #vector of q(3)_x;
y <- round(data.frame(a1, a2, a3), 4);
y1 <- data.frame(x, y);
y1;
```

The results are summarized in Table 1.20.

   It is to be noted that the data in this example are the same as in Example 1.4.6 and the results match with those in Example 1.4.6.

Table 1.20 Multiple decrement probabilities

Age x	$q_x^{(1)}$	$q_x^{(2)}$	$q_x^{(3)}$
55	0.0186	0.0258	0.1945
56	0.0201	0.0282	0.0974
57	0.0202	0.0305	0.1263
58	0.0212	0.0316	0.1165
59	0.0237	0.0348	0.1354

The multiple decrement model developed in this chapter provides a framework for studying many financial security systems. For example, life insurance policies frequently provide for special benefit if death occurs by accidental means or if the insured becomes disabled. In the next chapter, we discuss how basic methods used in calculating the actuarial present values of benefits, and hence premium calculations in a single decrement model get modified for a multiple decrement model. Another important application of multiple decrement models is in pension funding. It is discussed in Chaps. 3 and 4.

Key Terms   Associated single decrement tables, Cause of decrement random variable, Central rates of multiple decrement, Competing risks, Multiple decrement model, Multiple decrement table.

1.5 Exercises

1.1   A multiple decrement model with two causes of decrement has forces of decrement given by

$$\mu_x^{(1)}(t) = \frac{1}{100 - (x+t)} \quad \text{and} \quad \mu_x^{(2)}(t) = \frac{2}{100 - (x+t)}, \quad t < 100 - x.$$

Obtain the probability that

(i)   (40) survives for next 20 years.
(ii) (40) suffers decrement due to cause 1 in next 20 years.
(iii) (40) suffers decrement due to cause 2 in next 20 years.

1.2   Suppose that a multiple decrement model with 2 causes of decrement is specified by the following forces of decrement:

$$\mu_{x+t}^{(1)} = 0.0005 t a_1^x, \quad t \geq 0, \qquad \mu_{x+t}^{(2)} = 0.001 t a_2^x, \quad t \geq 0.$$

Suppose $a_1 = 1.02$ and $a_2 = 1.05$. For this model, obtain the joint distribution of $K(30)$ and $J(30)$. Also obtain the marginal distribution of $K(30)$ and $J(30)$ and their expected values.

Table 1.21 Decrement
probabilities

Curtate duration k	Probability of		
	Academic failure $q_k^{(1)}$	Withdrawal $q_k^{(2)}$	Completion $q_k^{(3)}$
0	0.15	0.25	0.00
1	0.10	0.20	0.00
2	0.05	0.15	0.00
3	0.00	0.10	0.00
4	0.00	0.00	1.00

1.3 A student entering a 4-semester master program quits the program due to three causes: academic failure during the program, $J = 1$, withdrawal due to various reasons, $J = 2$, and completing the program successfully at the end of fourth semester, $J = 3$. Table 1.21 specifies decrement probabilities which apply to students entering a 4-semester master program.
Suppose that 100 students are admitted to the program.

   (i) What is the probability that a student admitted to the program completes the program?
   (ii) What is the distribution of number of students completing the master program? State the underlying assumptions. Find its mean and variance. Use it to find approximate 3σ limits for the number of students completing the master program.
   (iii) What is the probability that a student admitted to the program will fail sometimes during the 4-semester program?
   (iv) What is the distribution of number of students who will fail sometimes during the 4-semester program? State the underlying assumptions. Find its mean and variance.
   (v) What is the probability that a student admitted to the program will withdraw sometimes during the 4-semester program?
   (vi) On the basis of given multiple decrement probabilities, construct a multiple decrement table displaying the expected number of decrements due to three causes at the end of each semester and the expected number of survivors at the beginning of each semester. Hence find the expected number of students completing the master program, the expected number of students who will fail sometime during the 4-semester program, and the expected number of withdrawals. Compare with the expected values obtained in (ii) and (iv). Using it, find the marginal distribution of mode of termination random variable J.
   (vii) Find the conditional distribution of J given that the student has terminated the program at the end of second semester.

1.4 Table 1.22 gives the probability of decrement due to two causes, second cause being the age-service retirement where 60 is the mandatory age of retirement.

Table 1.22 Decrement probabilities

x	$q_x^{(1)}$	$q_x^{(2)}$
50	0.0051	0.011
51	0.0054	0.023
52	0.0059	0.031
53	0.0065	0.042
54	0.0071	0.054
55	0.0075	0.062
56	0.0081	0.064
57	0.0092	0.076
58	0.0101	0.082
59	0.0112	0.093

Table 1.23 The number of survivors and number of deaths due to two causes

x	$l_x^{(\tau)}$	$d_x^{(1)}$	$d_x^{(2)}$
50	1000	8	5
51	987	13	9
52	965	10	9
53	946	13	10
54	923	13	15
55	895	14	12
56	869	16	18
57	835	17	20
58	798	20	23
59	755	22	28

Suppose that there are 1000 individuals of age 50 working in a company and they are subject to the decrement according to the probabilities given in Table 1.22.

(i) Find the expected number of individuals who retire at 60.
(ii) Find the expected number of decrements due to two causes in each of the year from 50 to 59.
(iii) Find the associated single decrement probabilities and the corresponding central rates of mortality under the assumptions to be stated.

1.5   Table 1.23 gives the number of survivors and number of deaths due to two causes.

Obtain the chance of decrement due to cause 1 and cause 2 and also the survival probability for all the ages. Compute the probabilities $_3p_{55}^{(\tau)}$, $_{3|}q_{53}^{(1)}$, and $_{3|}q_{56}^{(2)}$.

Table 1.24 The extract from a double-decrement table

x	$q_x^{(1)}$	$q_x^{(2)}$	$q_x^{(\tau)}$	$l_x^{(\tau)}$	$d_x^{(1)}$	$d_x^{(2)}$
60	–	–	0.0079	–	–	14
61	0.003	0.004	–	1600	–	–
62	–	–	–	–	4	–

Table 1.25 The extract from a double-decrement table

x	$l_x^{(\tau)}$	$d_x^{(1)}$	$d_x^{(2)}$
45	100	–	–
46	–	3	14
47	–	2	3

1.6   Calculate $_3q_{60}^{(1)}$ for a double-decrement model on the basis of information presented in Table 1.24.

1.7   Three forces of mortality for (x) are as given below:

$$\mu_{x+s}^{(1)} = 0.002 + 0.0021(x + s - 45)^{2.5},$$

$$\mu_{x+s}^{(2)} = 0.002 + 0.0005(x + s - 45)^3,$$

$$\mu_{x+s}^{(3)} = 0.002 + 0.00002(x + s - 45)^{3.3}, \quad s \geq 0, \ x \geq 45.$$

(i)   Obtain the associated single decrement table for ages 50 to 60.

(ii)   Under the assumption of uniformity for fractional ages, find the corresponding multiple decrement table.

(iii)   Using given forces of decrement, find the multiple decrement table and compare it with that obtained in (ii).

1.8   For a triple-decrement model, the values of $\mu_x^{(j)}$ for $j = 1, 2, 3$ are 0.06, 0.09, and 0.13, respectively.

(i)   Calculate $q_x^{(j)}$ and $q_x'^{(j)}$ for $j = 1, 2, 3$.

(ii)   Calculate the upper bound on $q_x'^{(j)}$ for $j = 1, 2, 3$, as derived in Example 1.4.1.

(iii)   Calculate $q_x'^{(j)}$ for $j = 1, 2, 3$ under the assumption of uniformity of deaths in a unit age interval.

(iv)   Compare the values in (iii) with the exact values and the upper bound calculated in (i) and (ii).

1.9   For a double-decrement model, it is given that the force of decrement is constant for each decrement over the year of age 45, 46, and 47, $q_{45}'^{(1)} = 0.02$, $q_{45}'^{(2)} = 0.08$, and further information is given in Table 1.25. Find $q_{47}'^{(1)}$.

Table 1.26 Absolute rates

Age x	$q_x^{\prime(1)}$	$q_x^{\prime(2)}$	$q_x^{\prime(3)}$
45	0.023	0.027	0.24
46	0.025	0.033	0.18
47	0.027	0.037	0.15
48	0.030	0.039	0.11
49	0.032	0.042	0.09
50	0.035	0.045	0.07

1.10 Given that decrement may be due to death, 1, disability, 2, or retirement, 3, use
     the constant force of decrement assumption to construct a multiple decrement
     table based on the following absolute rates (Table 1.26). Use the central rate
     bridge. State the underlying assumptions.

Chapter 2
Premiums and Reserves in Multiple Decrement Model

2.1 Introduction

A guiding principle in the determination of premiums for a variety of life insurance products is:

> Expected present value of inflow = Expected present value of outflow.

All the books listed in Sect. 1.1 of Chap. 1 thoroughly discuss the computation of premium for a variety of standard insurance products, for a single life and for a group in single decrement models. In this chapter we discuss its extension for multiple decrement models, when the benefit depends upon the mode of exit from the group of active insureds. Section 2.2 discusses how the multiple decrement model studied in Chap. 1 is useful to find the actuarial present value of benefit when it depends on the mode of decrement. Actuarial present value of the inflow to the insurance company, via premiums, does not depend on the mode of decrement. Hence, this part of the premium computations remains the same as for the single decrement model. When the two components of premiums are determined, premiums are calculated using the equivalence principle. Section 2.3 discusses the premium computations. In many life insurance products there is a provision of riders. For example, in whole life insurance the base policy specifies the benefit to be payable at the moment of death or at the end of year of death. Extra benefit will be payable if the death is due to a specific cause, such as an accident. The premium is then specified in two parts, one corresponding to the base policy and the extra premium corresponding to extra benefit. We will discuss computations of premiums in the presence of rider. Another important actuarial calculation is the reserve, that is, valuation of an insurance product at certain time points when the policy is in force. In Sect. 2.4 we illustrate the computation of reserve in the setup of multiple decrements.

S. Deshmukh, *Multiple Decrement Models in Insurance*,
DOI 10.1007/978-81-322-0659-0_2, © Springer India 2012

2.2 Actuarial Present Value of Benefit

Actuarial applications of multiple decrement models arise when the amount of benefit payment depends on the mode of exit from the group of active insureds. Our aim is to find the actuarial present value of the benefits in multiple decrement models, when the benefit is payable either at the moment of death or at the end of year of death. In these two approaches it depends upon the joint distribution of $T(x)$ and $J(x)$ or on the joint distribution of $K(x)$ and $J(x)$, respectively, and on the effective rate of interest. We assume that the rate of interest is deterministic and remains constant throughout the period of the policy. In practice the rate of interest fluctuates. If it is assumed to be deterministic but varying over certain time periods, then the actuarial present values of benefit or the annuity of premiums can be obtained on similar lines as that for constant rate, with different values of v or δ for the different time periods. The rate of interest is sometimes modeled as a random variable. We explore the modifications needed in Chap. 6.

Suppose that the underlying mortality model is the multiple decrement model with m causes of decrement. We consider the general setup in which benefit depends on the cause of decrement. This approach will be useful in the theory of pension funding in the next chapter. Suppose that $B^{(j)}_{x+t}$ denotes the value of a benefit at age $x + t$ for a decrement at that age by cause j. Then the actuarial present value of the benefit to be payable at the moment of death of (x), denoted in general by $\bar{A}$, is defined as $\bar{A} = E(B^{(J(x))}_{x+T(x)} v^{T(x)})$. We derive its expression in terms of basic functions as follows:

$$\bar{A} = E\left(B^{(J(x))}_{x+T(x)} v^{T(x)}\right)$$

$$= E_{J(x)}\left[E_{T(x)|J(x)}\left(B^{(J(x))}_{x+T(x)} v^{T(x)}\right)|J(x)\right]$$

$$= \sum_{j=1}^{m}\left[\int_{0}^{\infty} \left(B^{(j)}_{x+t} v^t f(t, j)/h_j\right) dt\right] h_j$$

$$= \sum_{j=1}^{m} \int_{0}^{\infty} B^{(j)}_{x+t} v^t \, {}_t p_x^{(\tau)} \mu_{x+t}^{(j)} \, dt. \tag{2.1}$$

If $m = 1$ and $B^{(j)}_{x+t} = 1$, $\bar{A}$ reduces to $\bar{A}_x$, the net single premium for whole life insurance with benefit payable at the moment of death. In general it is not easy to find these integrals. Some simplification can be obtained under certain assumptions. The most frequently made assumption is the assumption of uniformity for fractional ages. Suppose that we apply the uniform distribution assumption for each unit age interval in the jth integral, in (2.1). Then we have $T(x) = K(x) + U(x)$, where $U(x)$ has the uniform distribution on $(0, 1)$, and, further, $U(x)$ and $K(x)$ are independent random variables. Under this assumption, ${}_s p_{x+k}^{(\tau)} \mu_{x+k+s}^{(j)} = q_{x+k}^{(j)}$. With this, the jth integral in (2.1) reduces as follows:

$$\int_0^\infty B_{x+t}^{(j)} v^t \, {}_t p_x^{(\tau)} \mu_{x+t}^{(j)} \, dt = \sum_{k=0}^\infty \int_0^1 v^{k+s} B_{x+k+s}^{(j)} \, {}_{k+s} p_x^{(\tau)} \mu_{x+k+s}^{(j)} \, ds$$

$$= \sum_{k=0}^\infty v^{k+1} \, {}_k p_x^{(\tau)} q_{x+k}^{(j)} \int_0^1 B_{x+k+s}^{(j)} (1+i)^{1-s} \, ds$$

$$= \sum_{k=0}^\infty v^{k+1/2} \, {}_k p_x^{(\tau)} q_{x+k}^{(j)} B_{x+k+1/2}^{(j)},$$

where last step is obtained by the midpoint rule for the integral. The ${}_k p_x^{(\tau)}$ and $q_{x+k}^{(j)}$ values are available from the underlying multiple decrement table. Thus,

$$\bar{A} = \sum_{j=1}^m \sum_{k=0}^\infty v^{k+1/2} \, {}_k p_x^{(\tau)} q_{x+k}^{(j)} B_{x+k+1/2}^{(j)}$$

gives a practical formula for the evaluation of the integral.

We illustrate the computation for the n-year term insurance in the setup of a double indemnity provision in which the death benefit is doubled when death is caused by an accident. Let $J = 1$ for death by nonaccidental means, and $J = 2$ for death by accident and suppose that $B_{x+t}^{(1)} = 1$ and $B_{x+t}^{(2)} = 2$. We denote the net single premium for an n-year term insurance by $\bar{A}T$, and it is given by

$$\bar{A}T = \int_0^n v^t \, {}_t p_x^{(\tau)} \mu_{x+t}^{(1)} \, dt + 2 \int_0^n v^t \, {}_t p_x^{(\tau)} \mu_{x+t}^{(2)} \, dt.$$

We now assume that each decrement in the multiple decrement context has a uniform distribution in each year of age, and the first step is to break the expression into separate integrals for each of the years involved. The first integral can be expressed as

$$\int_0^n v^t \, {}_t p_x^{(\tau)} \mu_{x+t}^{(1)} \, dt = \sum_{k=0}^{n-1} v^k \, {}_k p_x^{(\tau)} \int_0^1 v^s \, {}_s p_{x+k}^{(\tau)} \mu_{x+k+s}^{(1)} \, ds.$$

Under the assumption of uniformity, we get

$$\int_0^n v^t \, {}_t p_x^{(\tau)} \mu_{x+t}^{(1)} \, dt = \sum_{k=0}^{n-1} v^{k+1} \, {}_k p_x^{(\tau)} q_{x+k}^{(1)} \int_0^1 (1+i)^{1-s} \, ds = \frac{i}{\delta} \sum_{k=0}^{n-1} v^{k+1} \, {}_k p_x^{(\tau)} q_{x+k}^{(1)}.$$

Applying a similar argument for the second integral and combining, we get

$$\bar{A}T = \frac{i}{\delta} \left[\sum_{k=0}^{n-1} v^{k+1} \, {}_k p_x^{(\tau)} \left(q_{x+k}^{(1)} + 2q_{x+k}^{(2)} \right) \right]$$

$$= \frac{i}{\delta} \sum_{k=0}^{n-1} v^{k+1} \, {}_k p_x^{(\tau)} q_{x+k}^{(2)} + \frac{i}{\delta} \sum_{k=0}^{n-1} v^{k+1} \, {}_k p_x^{(\tau)} q_{x+k}^{(\tau)}$$

$$= \bar{A}_{x:\overline{n}|}^{1(2)} + \bar{A}_{x:\overline{n}|}^{1},$$

where $\bar{A}_{x:\overline{n}|}^{1(2)}$ is the net single premium for a term insurance of 1 covering death from accidental means, and $\bar{A}_{x:\overline{n}|}^{1}$ is the net single premium for a term insurance of 1 covering death from all causes. The next example illustrates the computation for specified mortality pattern and for specified benefit values.

Example 2.2.1 A whole life insurance of 10000 payable at the moment of death of (x) includes a double-indemnity provision. This provision pays an additional death benefit of 10000 during the first 20 years if death is by accidental means. It is given that $\delta = 0.05$, $\mu_{x+t}^{(\tau)} = 0.005$ for $t \geq 0$, and $\mu_{x+t}^{(1)} = 0.001$ for $t \geq 0$, where $\mu_{x+t}^{(1)}$ is the force of decrement due to death by accidental means. Calculate the net single premium for this insurance.

Solution From the given information, the net single premium is the actuarial present value of the benefit of 10000 in whole life insurance plus the actuarial present value of the benefit of 10000 in 20-year term insurance if death is due to accident. Thus it is given by the expression

$$10000 \left[\int_0^\infty e^{-\delta t} \, {}_t p_x^{(\tau)} \mu_{x+t}^{(\tau)} \, dt + \int_0^{20} e^{-\delta t} \, {}_t p_x^{(\tau)} \mu_{x+t}^{(1)} \, dt \right]$$

$$= 10000 \left[\int_0^\infty e^{-0.05t} e^{-0.005t} 0.005 \, dt + \int_0^{20} e^{-0.05t} e^{-0.005t} 0.001 \, dt \right]$$

$$= 10000 \left[\frac{0.005}{0.055} + \frac{0.001}{0.055} (1 - e^{-1.1}) \right] = 1030.$$

Example 2.2.2 For a 20-year term insurance issued to (30), the following information is given.

(i) $\mu_{30+t}^{(1)} = 0.0005t$, where (1) represents death by accidental means.
(ii) $\mu_{30+t}^{(2)} = 0.0025t$, where (2) represents death by other means.
(iii)   The benefit is 2000 units if death occurs by accidental means and 1000 units if death occurs by other means.
(iv)   Benefits are payable at the moment of death.

Taking $\delta = 0.06$, find the purchasing price of this insurance.

Solution From the given information we have

$$\mu_{x+t}^{(\tau)} = 0.003t \quad \Rightarrow \quad {}_t p_x^{(\tau)} = \exp\left[-\int_0^t \mu_{x+s}^{(\tau)} \, ds \right] = e^{-0.0015t^2}.$$

The actuarial present value of accidental death benefit is

$$\int_0^{20} B^{(1)}_{30+t} e^{-\delta t} \, {}_t p^{(\tau)}_{30} \, \mu^{(1)}_{30+t} \, dt = 2000 \int_0^{20} e^{-0.06t} e^{-0.0015t^2} (0.0005t) \, dt.$$

To workout this integral, we complete the square in the exponent of e and substitute $0.0015(t + 20)^2 = u$. Further we use the incomplete gamma function to find the value of the integral. Thus, we have

$$2000 \int_0^{20} e^{-0.06t} e^{-0.0015t^2} (0.0005t) \, dt$$

$$= \int_0^{20} (t + 20 - 20) e^{-0.0015(t^2 + 40t + 400 - 400)} \, dt$$

$$= \frac{e^{0.6}}{0.003} \left\{ \left[\int_{0.6}^{2.4} e^{-u} \, du - 20(0.0015)^{0.5} \Gamma(0.5) \right. \right.$$

$$\left. \left. \times \left[pgamma(2.4, 0.5, 1) - pgamma(0.6, 0.5, 1) \right] \right] \right\}$$

$$= 74.05.$$

The actuarial present value of other death benefit is calculated using the similar procedure adopted for the actuarial present value of accidental death benefit and is given by

$$1000 \int_0^{20} B^{(2)}_{30+t} e^{-\delta t} \, {}_t p^{(\tau)}_{30} \, \mu^{(2)}_{30+t} \, dt = 1000 \int_0^{20} e^{-0.06t} e^{-0.0015t^2} (0.0025t) \, dt$$

$$= 185.12.$$

The actuarial present value of death benefit, when death may be due to any cause, is 259.17, which is the purchasing price of the insurance product.

We have seen how to find the actuarial present value of the benefits in multiple decrement models, when the benefit is payable at the moment of death. When the benefit is payable at the end of year of death, we use the joint distribution of $K(x)$ and $J(x)$ instead of the joint distribution of $T(x)$ and $J(x)$. Let $\{p(k, j), k = 0, 1, \ldots, j = 1, 2, \ldots, m\}$ denote the joint probability mass function of $K(x)$ and $J(x)$. The actuarial present value of the benefit to be payable at the end of the year of death in whole life insurance for multiple decrement model, to be in general denoted by A, is given by

$$A = E\left(B^{(J(x))}_{x+K(x)} v^{K(x)+1}\right) = \sum_{j=1}^m \sum_{k=0}^\infty v^{k+1} B^{(j)}_{x+k} p(k, j).$$

Similarly, the actuarial present value of the benefit to be payable at the end of the year of death in n-year term insurance for multiple decrement model is given by

$$AT = \sum_{j=1}^{m} \sum_{k=0}^{n-1} v^{k+1} B_{x+k}^{(j)} P(k, j).$$

Actuarial present values for other insurance products are defined analogously.

In the next section we discuss the procedure for premium computation and illustrate it with examples.

2.3 Computation of Premiums

The actuarial present value of the benefit to be payable at the moment of death or at the end of year of death, is one of the two components in premium calculation. The other component in premium calculation is the expected present value of inflow to the insurance company via premiums. These computations remain exactly the same as in the setup of single decrement model and are not changed in view of various modes of decrement. Suppose that the premiums are paid as continuous whole life annuity at the rate of P per annum. Then the actuarial present value of the premiums is $P\bar{a}_x$, and $\bar{a}_x$ is given by

$$\bar{a}_x = \int_0^{\infty} v^t \, {}_t p_x^{(\tau)} \, dt.$$

Suppose that the premiums are paid as continuous n-year temporary life annuity at the rate of P per annum. Then the actuarial present value of the premiums is $P\bar{a}_{x:\bar{n}|}$, and $\bar{a}_{x:\bar{n}|}$ is given by

$$\bar{a}_{x:\bar{n}|} = \int_0^{n} v^t \, {}_t p_x^{(\tau)} \, dt.$$

If the premiums are paid as discrete whole life annuity due at the rate of P per annum, then the actuarial present value of the premiums is $P\ddot{a}_x$, and $\ddot{a}_x$ is given by

$$\ddot{a}_x = \sum_{k=0}^{\infty} v^k \, {}_k p_x^{(\tau)}.$$

Suppose that the premiums are payable as a discrete n-year temporary life annuity due at the rate of P per annum. Then the actuarial present value of the premiums is $P\ddot{a}_{x:\bar{n}|}$, and $\ddot{a}_{x:\bar{n}|}$ is given by

$$\ddot{a}_{x:\bar{n}|} = \sum_{k=0}^{n-1} v^k \, {}_k p_x^{(\tau)}.$$

Thus, once we have knowledge about the survival function $_t p_x^{(\tau)}$, we can find out the actuarial present value of the premiums for various modes of premium payments, such as continuous premium and discrete premium. With the equivalence principle, premium is then obtained as

$$\text{Premium} = \frac{\text{Actuarial Present Value of Benefits}}{\text{Actuarial Present Value of Premium Annuity}}.$$

The following examples illustrate how the premium computations can be done once the multiple decrement model is specified in terms of forces of mortality.

Example 2.3.1 For the insurance product, mortality pattern, and force of interest as specified in Example 2.2.1, find the premium payable as

  (i) the whole life continuous annuity,
 (ii) whole life annuity due, and
(iii) 10-year temporary continuous life annuity.

Solution For the given insurance product, mortality pattern, and force of interest, we have obtained the actuarial present value of the benefit payable at the moment of death. It is Rs 1030. To find the premium payable as (i) the whole life continuous annuity, (ii) whole life annuity due, and (iii) 10-year temporary continuous life annuity, we compute $\bar{a}_x$, $\ddot{a}_x$, and $\bar{a}_{x:\overline{10|}}$, respectively, for the given mortality and interest pattern. By definition,

$$\bar{a}_x = \int_0^\infty v^t \, _t p_x^{(\tau)} \, dt = \int_0^\infty e^{-0.05t} e^{-0.005t} \, dt = 18.18182,$$

$$\ddot{a}_x = \sum_{k=0}^\infty v^k \, _k p_x^{(\tau)} = \sum_{k=0}^\infty e^{-0.05k} e^{-0.005k} = \left(1 - e^{-0.055}\right)^{-1} = 18.6864 > \bar{a}_x,$$

$$\bar{a}_{x:\overline{10|}} = \int_0^{10} v^t \, _t p_x^{(\tau)} \, dt = \int_0^{10} e^{-0.05t} e^{-0.005t} = 7.6918.$$

Hence, (i) the premium payable as the whole life continuous annuity is $1030/18.18182 = 56.65$, (ii) the premium payable as the whole life annuity due is $1030/18.6864 = 55.12$, and (iii) the premium payable as the 10-year temporary continuous life annuity is $1030/7.6918 = 133.91$. It is to be noted that the premium payable as the 10-year temporary continuous life annuity is highest among these three modes as the premium paying period is limited. Further, the premium payable as the whole life continuous annuity is slightly higher than that payable as the whole life annuity due.

Example 2.3.2 A multiple decrement model with two causes of decrement is given below in terms of the forces of decrement as

$$\mu_{x+t}^{(1)} = BC^{x+t}, \quad t \geq 0, \qquad \mu_{x+t}^{(2)} = A, \quad t \geq 0, \ A \geq 0, \ B \geq 0, \ C \geq 1.$$

Suppose $A = 0.0008$, $B = 0.00011$, and $C = 1.095$. Further, the force of interest is $\delta = 0.05$. The benefit to be payable at the moment of death is specified as 1000 units if death is due to cause 1 and 2000 units if the death is due to cause 2.

(i) Find the actuarial present value of the benefit payable to (x) for $x = 30, 40, 50,$ and 60 for the whole life insurance.
(ii) Find the premium payable as the continuous whole life annuity by (x) for $x = 30, 40, 50,$ and 60 for the whole life insurance. Decompose the total premium according to two causes of decrement.
(iii) Find the premium payable as the continuous n-year temporary life annuity by (30) for the whole life insurance, for $n = 1, 2, \ldots, 10$. Decompose the total premium according to two causes of decrement.
(iv) Find the premium payable as the continuous n-year temporary life annuity by (30), for n-year term insurance, for $n = 1, 2, \ldots, 10$.

Solution For the given model, $\mu_{x+s}^{(\tau)} = \mu_{x+s}^{(1)} + \mu_{x+s}^{(2)} = BC^{x+s} + A$. We have derived in Example 1.2.2 the probability density function of T as

$$g(t) = {}_t p_x^{(\tau)} \mu_{x+t}^{(\tau)} = \exp\left[-\left(At + mC^x(C^t - 1)\right)\right]\left(A + BC^{x+t}\right) \quad \text{if } t \geq 0$$

and the joint distribution of T and J as

$$f(t, j) = \begin{cases} \exp[-(At + mC^x(C^t - 1))](BC^{x+t}) & \text{if } t \geq 0, \; j = 1, \\ \exp[-(At + mC^x(C^t - 1))](A) & \text{if } t \geq 0, \; j = 2. \end{cases}$$

Let mC^x denote by α_x. The actuarial present value of the benefit to be payable at the moment of death of (x), denoted by $\bar{A}$, is given by

$$\bar{A} = \sum_{j=1}^{2} \int_0^\infty B_{x+t}^{(j)} v^t \, {}_t p_x^{(\tau)} \mu_{x+t}^{(j)} \, dt$$

$$= 1000 BC^x e^{\alpha_x} \int_0^\infty e^{-\delta t - At - \alpha_x C^t} C^t \, dt + 2000 A e^{\alpha_x} \int_0^\infty e^{-\delta t - At - \alpha_x C^t} \, dt$$

$$= 1000 BC^x e^{\alpha_x} \int_0^\infty e^{-\delta_1 t - \alpha_x C^t} C^t \, dt + 2000 A e^{\alpha_x} \int_0^\infty e^{-\delta_1 t - \alpha_x C^t} \, dt,$$

where $\delta_1 = \delta + A$. Then, substituting $C^t = y$, we get $e^{-\delta_1 t} = y^{-\delta_1/\log C}$, $C^t \log C \, dt = dy$, and the range of integration is from 1 to ∞. Suppose $(-\delta_1/\log C) + 1 = \lambda_1$. Then, as in Example 1.2.2, the first integral in $\bar{A}$ simplifies to $1000 e^{\alpha_x} \Gamma(\lambda_1) \alpha_x^{1-\lambda_1} P[W_1 \geq 1]$, where W_1 follows the gamma distribution with shape parameter λ_1 and scale parameter α_x, provided that $\lambda_1 > 0$. Now, using the fact that $g(t)$ is a density function, we get

$$1 = \int_0^\infty g(t) \, dt = \int_0^\infty \exp\left[-\left(At + mC^x(C^t - 1)\right)\right]\left(A + BC^{x+t}\right) dt.$$

Hence,

$$Ae^{\alpha_x} \int_0^\infty \exp\left[-\left(At + \alpha_x C^t\right) dt\right] = 1 - e^{\alpha_x} \Gamma(\lambda) \alpha_x^{1-\lambda} P[W \geq 1],$$

where $\lambda = (-A/\log C) + 1$, and W has the gamma distribution with shape parameter λ and scale parameter α_x. Using this approach, the second term in $\bar{A}$ can be expressed as

$$\frac{2000A[1 - e^{\alpha_x} \Gamma(\lambda_1) \alpha_x^{1-\lambda_1} P[W_1 \geq 1]]}{\delta_1}.$$

Hence, adding the two components of $\bar{A}$, we get

$$\bar{A} = \frac{2000A + 1000(\delta - A)[e^{\alpha_x} \Gamma(\lambda_1) \alpha_x^{1-\lambda_1} P[W_1 \geq 1]]}{A + \delta}.$$

Now the premium is payable as the continuous whole life annuity. So we need to find the actuarial present value of the premiums. It is given by $P\bar{a}_x$, and $\bar{a}_x$ is given by

$$\bar{a}_x = \int_0^\infty v^t\,_t p_x^{(\tau)}\, dt$$

$$= e^{\alpha_x} \int_0^\infty e^{-\delta t - At - \alpha_x C^t}\, dt = e^{\alpha_x} \int_0^\infty e^{-\delta_1 t - \alpha_x C^t}\, dt$$

$$= \frac{1 - e^{\alpha_x} \Gamma(\lambda_1) \alpha_x^{1-\lambda_1} P[W_1 \geq 1]}{\delta_1}.$$

The last equality follows using similar arguments as in the second integral of $\bar{A}$.

Suppose that the premium is payable as the continuous n-year temporary life annuity. The actuarial present value of the premiums is given by $P\bar{a}_{x:\bar{n}|}$, and $\bar{a}_{x:\bar{n}|}$ is given by

$$\bar{a}_{x:\bar{n}|} = \int_0^n v^t\,_t p_x^{(\tau)}\, dt$$

$$= e^{\alpha_x} \int_0^n e^{-\delta t - At - \alpha_x C^t}\, dt = e^{\alpha_x} \int_0^n e^{-\delta_1 t - \alpha_x C^t}\, dt,$$

where $\delta_1 = A + \delta$. To find this integral, we proceed as follows. We know that, for the two decrement-model in this example,

$$\int_0^n g(t)\, dt = P[T(x) \leq n] = {}_n q_x^{(\tau)} = 1 - {}_n p_x^{(\tau)} = 1 - \exp[-An - \alpha_x C^n + \alpha_x].$$

On the other hand, $\int_0^n g(t)\,dt$ can also be expressed as follows.:

$$
\int_0^n g(t)\,dt = \int_0^n {}_t p_x^{(\tau)}\left(A + BC^{x+t}\right)dt
$$

$$
= A \int_0^n {}_t p_x^{(\tau)}\,dt + \int_0^n {}_t p_x^{(\tau)}\left(BC^{x+t}\right)dt
$$

$$
= A e^{\alpha_x} \int_0^n \exp\left(-At - \alpha_x C^t\right)dt
$$

$$
+ BC^x e^{\alpha_x} \int_0^n \exp\left(-At - \alpha_x C^t\right)C^t\,dt.
$$

The second integral in the above equation can be evaluated as $e^{\alpha_x}\Gamma(\lambda)\alpha_x^{1-\lambda}P[1 \leq W \leq C^n]$, where $(-\delta/\log C) + 1 = \lambda$, and W follows the gamma distribution with shape parameter λ and scale parameter α_x. Hence the first integral $A\int_0^n {}_t p_x^{(\tau)}\,dt$ in the above equation is given by

$$
A e^{\alpha_x} \int_0^n \exp\left(-At - \alpha_x C^t\right)dt = 1 - \exp\left[-An - \alpha_x C^n + \alpha_x\right]
$$

$$
- e^{\alpha_x}\Gamma(\lambda)\alpha_x^{1-\lambda}P\left[1 \leq W \leq C^n\right].
$$

We use this relation to write $\bar{a}_{x:\bar{n}|}$ as

$$
\bar{a}_{x:\bar{n}|} = \left\{1 - \exp\left[-\delta_1 n - \alpha_x C^n + \alpha_x\right] - e^{\alpha_x}\Gamma(\lambda_1)\alpha_x^{1-\lambda_1}P\left[1 \leq W_1 \leq C^n\right]\right\}/\delta_1,
$$

where $\delta_1 = A + \delta$, $(-\delta_1/\log C) + 1 = \lambda_1$, and W_1 follows the gamma distribution with shape parameter λ_1 and scale parameter α_x, provided that $\lambda_1 > 0$.

The annual premium payable as the whole life annuity for benefit of 1000 units if death is due to cause 1 and 2000 units if the death is due to cause 2 is then given by $P = \frac{\bar{A}}{\bar{a}_x}$. To decompose the total premium according to two causes of decrement, we divide two terms in $\bar{A}$ separately by $\bar{a}_x$. Let $P_x^{(1)}$ and $P_x^{(2)}$ denote the premiums corresponding to two causes of decrement; then these are given by

$$
P_x^{(1)} = \frac{1000 e^{\alpha_x}\Gamma(\lambda_1)\alpha_x^{1-\lambda_1}P[W_1 \geq 1]}{\bar{a}_x} \quad \text{and}
$$

$$
P_x^{(2)} = \frac{2000A[1 - e^{\alpha_x}\Gamma(\lambda_1)\alpha_x^{1-\lambda_1}P[W_1 \geq 1]]}{\delta_1 \bar{a}_x}.
$$

Substituting the expression for $\bar{a}_x$ into $P_x^{(2)}$, we get

$$
P_x^{(2)} = 2000A.
$$

It is constant and does not depend on x. It seems reasonable as the force of decrement due to cause 2 is free from x. Premiums payable as the continuous n-year temporary life annuity for the whole life insurance are computed on similar lines, denominator being $\bar{a}_{x:\bar{n}|}$ instead of $\bar{a}_x$.

To find premiums for the n-year term, we have to first find the actuarial present value of the benefit for this product. The actuarial present value of the benefit to be payable at the moment of death of (x), in the n-year term insurance, denoted by $\bar{A}T$, is given by

$$\bar{A}T = \sum_{j=1}^{2} \int_0^n B_{x+t}^{(j)} v^t \, {}_t p_x^{(\tau)} \mu_{x+t}^{(j)} \, dt$$

$$= 1000BC^x e^{\alpha_x} \int_0^n e^{-\delta t - At - \alpha_x C^t} C^t dt + 2000 A e^{\alpha_x} \int_0^n e^{-\delta t - At - \alpha_x C^t} dt$$

$$= 1000BC^x e^{\alpha_x} \int_0^n e^{-\delta_1 t - \alpha_x t C^t} C^t dt + 2000 A e^{\alpha_x} \int_0^n e^{-\delta_1 t - \alpha_x C^t} dt,$$

where $\delta_1 = \delta + A$. As discussed above, the first integral in the above equation can be expressed as $1000 e^{\alpha_x} \Gamma(\lambda_1) \alpha_x^{1-\lambda_1} P[1 \le W_1 \le C^n]$, where $(-\delta_1/\log C) + 1 = \lambda_1$, and W_1 follows the gamma distribution with shape parameter λ_1 and scale parameter α_x. As in the expression of $\bar{a}_{x:\bar{n}|}$, the second integral in $\bar{A}T$ can be expressed as

$$2000(A/\delta_1)\left\{1 - \exp\left[-\delta_1 n - \alpha_x C^n + \alpha_x\right] - e^{\alpha_x} \Gamma(\lambda_1) \alpha_x^{1-\lambda_1} P\left[1 \le W_1 \le C^n\right]\right\}$$

$$= 2000 A \bar{a}_{x:\bar{n}|}.$$

Thus, $\bar{A}T$ is given by

$$\bar{A}T = 1000 e^{\alpha_x} \Gamma(\lambda_1) \alpha_x^{1-\lambda_1} P\left[1 \le W_1 \le C^n\right] + 2000 A \bar{a}_{x:\bar{n}|}.$$

We have already derived the expression for $\bar{a}_{x:\bar{n}|}$. Once we have expressions for these actuarial present values, we can compute premiums using the equivalence principle. Thus,

$$P_{x:\bar{n}|}^{1(1)} = \frac{1000 e^{\alpha_x} \Gamma(\lambda_1) \alpha_x^{1-\lambda_1} P[1 \le W_1 \le C^n]}{\bar{a}_{x:\bar{n}|}} \quad \text{and}$$

$$P_{x:\bar{n}|}^{1(2)} = \frac{2000 A \bar{a}_{x:\bar{n}|}}{\bar{a}_{x:\bar{n}|}} = 2000 A,$$

similar to $P_x^{(2)}$.

We compute all these quantities using the following R commands:

```
a1 <- 0.0008   # A;
b <- 0.00011   # B;
a <- 1.095   # C;
m <- b/log(a, base=exp(1));
e <- exp(1);
```

Table 2.1 Premium for whole life insurance

Age x	$1000\bar{A}$	$\bar{a}_x$	$P_x^{(1)}$	$P_x^{(2)}$	P_x
30	202.77	16.2039	10.91	1.60	12.51
40	290.39	14.4229	18.53	1.60	20.13
50	406.68	12.0593	32.12	1.60	33.72
60	545.70	9.2338	57.50	1.60	59.10

```
del <- 0.05;
f <- a1+del;
p <- (-f/log(a, base=exp(1)))+1   # λ1;
x <- c(30, 40, 50, 60);
j <- m*a^x  # αx;
q1 <- e^j*gamma(p)*(j^(1-p))*(1-pgamma(1, p, j))
                                  # first term in Ā;
q2 <- (a1/f)*(1-q1)  # second integral in Ā;
q3 <- 1000*q1+2000*q2  # Ā;
q4 <- (1-q1)/f  # āx;
p1 <- 1000*q1/q4  # premium corresponding to cause 1;
p2 <- 1000*2*q2/q4  # premium corresponding to cause 2;
p3 <- p1+p2  # premium ;
d <- round(data.frame(q3, q4, p1, p2, p3), 4);
d1 <- data.frame(x, d);
d1   # Table 2.1;
```

The actuarial present values of benefit and annuity and the corresponding premiums for four ages are reported in Table 2.1. It is to be noted that $P_x^{(2)} = 2000A = 1.6$. Suppose that the premium are payable as the n-year temporary life annuity by (30). Then the following commands are added after the command for p in the above set to obtain the premiums:

```
x <- 30;
n <- 1:10;
j <- m*a^x;
q1 <- e^j*gamma(p)*(j^(1-p))*(1-pgamma(1, p, j));
q2 <- (a1/f)*(1-q1);
q <- e^j*gamma(p)*(j^(1-p))*(pgamma(a^n, p, j)
     -pgamma(1, p, j));
q4 <- (1-q-e^(j-f*n-j*a^n))/f   # āx:n̄|;
p1 <- 1000*q1/q4;
p2 <- 1000*2*q2/q4;
p3 <- p1+p2;
d <- round(data.frame(p1, p2, p3), 2);
d1 <- data.frame(n, d);
d1   # Table 2.2;
```

Table 2.2 n-year premium payment

n	$_nP_{30}^{(1)}$	$_nP_{30}^{(2)}$	$_nP_{30}$
1	181.53	26.61	208.14
2	93.15	13.66	106.81
3	63.72	9.34	73.07
4	49.03	7.19	56.22
5	40.24	5.90	46.14
6	34.39	5.04	39.43
7	30.23	4.43	34.66
8	27.12	3.98	31.09
9	24.71	3.62	28.33
10	22.79	3.34	26.13

Table 2.3 Premium for n-year term insurance

n	$P_{30:\overline{n}|}^{1(1)}$	$P_{30:\overline{n}|}^{1(2)}$	$P_{30:\overline{n}|}^{1}$
1	1.75	1.6	3.35
2	1.83	1.6	3.43
3	1.92	1.6	3.52
4	2.01	1.6	3.61
5	2.10	1.6	3.70
6	2.19	1.6	3.79
7	2.29	1.6	3.89
8	2.40	1.6	4.00
9	2.51	1.6	4.11
10	2.62	1.6	4.22

Premiums payable as the n-year temporary annuity for the whole life insurance are reported in Table 2.2.

To compute the premium for the n-year term insurance, in the above set of commands, we add the following R commands:

```
q3 <- 2*a1*(1-q-e^(j-f*n-j*a^n))/f;
pr1 <- 1000*q/q4;
pr2 <- 1000*q3/q4;
pr3 <- pr1+pr2;
d <- round(data.frame(pr1, pr2, pr3), 2);
d1 <- data.frame(n, d);
d1   #Table 2.3;
```

Table 2.3 gives the premiums for the n-year term insurance.

It is to be noted that, as usual, the premium in the n-year term insurance is always less than that for the whole life insurance.

In the next example, we use the same multiple decrement model used in Example 2.3.2; however the benefit functions are taken as increasing functions and also dependent on age.

Example 2.3.3 A multiple decrement model with two causes of decrement is given below in terms of the forces of decrement as

$$\mu_{x+t}^{(1)} = BC^{x+t}, \quad t \geq 0, \qquad \mu_{x+t}^{(2)} = A, \quad t \geq 0, \ A \geq 0, \ B \geq 0, \ C \geq 1.$$

Suppose $A = 0.0008$, $B = 0.00011$, and $C = 1.095$. Further, the force of interest is $\delta = 0.05$. The benefit to be payable at the moment of death is specified as $1000e^{b_1(x+t)}$ units if death is due to cause 1 and $1000e^{b_2(x+t)}$ units if the death is due to cause 2, where $b_1 = 0.02$, $b_2 = 0.03$.

(i) Find the actuarial present value of the benefit payable to (x) for $x = 30, 40, 50$, and 60 for the whole life insurance.
(ii) Find the premium payable as the continuous life annuity by (x) for $x = 30, 40, 50$, and 60 for the whole life insurance. Decompose the total premium according to two causes of decrement.
(iii) Find the premium payable as the continuous n-year temporary life annuity by (30) for the whole life insurance, for $n = 1, 2, \ldots, 10$. Decompose the total premium according to two causes of decrement.
(iv) Find the premium payable as the continuous n-year temporary life annuity by (30), for n-year term insurance, for $n = 1, 2, \ldots, 10$.

Solution As in Example 2.3.2, for this multiple decrement model, $\mu_{x+s}^{(\tau)} = \mu_{x+s}^{(1)} + \mu_{x+s}^{(2)} = BC^{x+s} + A$. The probability density function of T is

$$g(t) = {}_t p_x^{(\tau)} \mu_{x+t}^{(\tau)} = \exp\left[-\left(At + mC^x(C^t - 1)\right)\right]\left(A + BC^{x+t}\right) \quad \text{if } t \geq 0,$$

and the joint distribution of T and J is specified by

$$f(t, j) = \begin{cases} \exp[-(At + mC^x(C^t - 1))](BC^{x+t}) & \text{if } t \geq 0, \ j = 1, \\ \exp[-(At + mC^x(C^t - 1))](A) & \text{if } t \geq 0, \ j = 2. \end{cases}$$

Let mC^x be denoted by α_x. The actuarial present value of the benefit, omitting 1000, to be payable at the moment of death of (x), denoted by $\bar{A}$, is given by

$$\bar{A} = \sum_{j=1}^{2} \int_0^\infty B_{x+t}^{(j)} v^t \, {}_t p_x^{(\tau)} \mu_{x+t}^{(j)} \, dt$$

$$= \int_0^\infty e^{b_1(x+t)} e^{-\delta t} \, {}_t p_x^{(\tau)} \mu_{x+t}^{(1)} \, dt + \int_0^\infty e^{b_2(x+t)} e^{-\delta t} \, {}_t p_x^{(\tau)} \mu_{x+t}^{(2)} \, dt$$

$$= BC^x e^{b_1 x + \alpha_x} \int_0^\infty e^{b_1 t - \delta t - At - \alpha_x C^t} C^t \, dt + A e^{b_2 x + \alpha_x} \int_0^\infty e^{b_2 t - \delta t - At - \alpha_x C^t} \, dt$$

$$= BC^x e^{b_1 x + \alpha_x} \int_0^\infty e^{-\delta_2 t - \alpha_x C^t} C^t \, dt + A e^{b_2 x + \alpha_x} \int_0^\infty e^{-\delta_3 t - \alpha_x C^t} \, dt,$$

where $\delta_2 = \delta + A - b_1$ and $\delta_3 = \delta + A - b_2$. Suppose $(-\delta_2/\log C) + 1 = \lambda_2$ and $(-\delta_3/\log C) + 1 = \lambda_3$. If λ_1 and λ_2 are positive, then, as in Example 2.3.1, the first integral in $\bar{A}$ simplifies to $e^{b_1 x + \alpha_x} \Gamma(\lambda_2) \alpha_x^{1-\lambda_2} P[W_2 \geq 1]$, where W_2 follows the gamma distribution with shape parameter λ_2 and scale parameter α_x. Similarly, as in Example 2.3.1, the second integral in $\bar{A}$ simplifies to

$$\frac{e^{b_2 x} A\{1 - e^{\alpha_x} \Gamma(\lambda_3) \alpha_x^{1-\lambda_3} P[W_3 \geq 1]\}}{\delta_3},$$

where W_3 follows the gamma distribution with shape parameter λ_3 and scale parameter α_x.

Hence, $\bar{A}$ is given by

$$e^{b_1 x + \alpha_x} \Gamma(\lambda_2) \alpha_x^{1-\lambda_2} P[W_2 \geq 1] + \frac{e^{b_2 x} A\{1 - e^{\alpha_x} \Gamma(\lambda_3) \alpha_x^{1-\lambda_3} P[W_3 \geq 1]\}}{\delta_3}.$$

Expressions for $\bar{a}_x$ and $\bar{a}_{x:\overline{n}|}$ will remain the same as in Example 2.3.1. Once we have both components of premium calculations, we can find the premiums.

To find the premium for the n-year term insurance, we have to first find the actuarial present value of the benefit for this insurance. The actuarial present value of the benefit to be payable at the moment of death of (x), in the n-year term insurance, denoted by $\bar{A}T$, can be obtained on similar lines as in $\bar{A}T$ in Example 2.3.2. It is given by

$$\bar{A}T = \sum_{j=1}^2 \int_0^n B_{x+t}^{(j)} v^t \, {}_t p_x^{(\tau)} \mu_{x+t}^{(j)} \, dt$$

$$= \int_0^n e^{b_1(x+t)} e^{-\delta t} \, {}_t p_x^{(\tau)} \mu_{x+t}^{(1)} \, dt + \int_0^n e^{b_2(x+t)} e^{-\delta t} \, {}_t p_x^{(\tau)} \mu_{x+t}^{(2)} \, dt$$

$$= BC^x e^{b_1 x + \alpha_x} \int_0^n e^{-\delta_2 t - \alpha_x t C^t} C^t \, dt + A e^{b_2 x + \alpha_x} \int_0^n e^{-\delta_3 t - \alpha_x C^t} \, dt$$

$$= e^{\alpha_x + b_1 x} \Gamma(\lambda_2) \alpha_x^{1-\lambda_2} P[1 \leq W_2 \leq C^n]$$

$$+ e^{b_2 x} \frac{A}{\delta_3} \{1 - \exp[-\delta_3 n - \alpha_x C^n + \alpha_x]$$

$$- e^{\alpha_x} \Gamma(\lambda_3) \alpha_x^{1-\lambda_3} P[1 \leq W_3 \leq C^n]\}.$$

We have already derived the expression for $\bar{a}_{x:\overline{n}|}$. Once we have expressions for these actuarial present values, we can compute the premiums using equivalence principle. The following R commands compute all these functions:

Table 2.4 Premiums for varying benefits

Age x	$\bar{A}$	$\bar{a}_x$	$P_x^{(1)}$	$P_x^{(2)}$	P_x
30	0.6527	16.2039	37.15	3.12	40.28
40	1.0142	14.4229	66.40	3.92	70.32
50	1.5441	12.0593	123.16	4.89	128.04
60	2.2790	9.2338	240.71	6.10	246.81

```
a1 <- 0.0008    # A;
b <- 0.00011    # B;
a <- 1.095    # C;
m <- b/log(a, base=exp(1));
e <- exp(1);
del <- 0.05;
b1 <- 0.02;
b2 <- 0.03;
f <- a1+del;
p <- (-f/log(a, base=exp(1)))+1    # λ₁;
f1 <- a1+del-b1;
p1 <- (-f1/log(a, base=exp(1)))+1    # λ₂;
f2 <- a1+del-b2;
p2 <- (-f2/log(a, base=exp(1)))+1    # λ₃;
x <- c(30, 40, 50, 60);
j <- m*a^x    # αₓ;
q <- e^j*gamma(p)*(j^(1-p))*(1-pgamma(1, p, j));
q1 <- e^j*gamma(p1)*(j^(1-p1))*(1-pgamma(1, p1, j));
q2 <- e^j*gamma(p2)*(j^(1-p2))*(1-pgamma(1, p2, j));
q3 <- e^(b2*x)*a1*(1-q2)/f2   # second term in Ā;
q4 <- (1-q)/f;   # āₓ;
q5 <- e^(b1*x)*q1+q3    # Ā;
pr1 <- 1000*e^(b1*x)*q1/q4
                     # premium corresponding to cause 1;
pr2 <- 1000*q3/q4   # premium corresponding to cause 2;
pr3 <- pr1+pr2;
d <- round(data.frame(q5, q4, pr1, pr2, pr3), 4);
d1 <- data.frame(x, d);
d1    # Table 2.4;
```

Premiums are reported in Table 2.4.

It is to be noted that the premiums are higher as compared to those reported in Table 2.1, as here the benefit function is an increasing function for both modes of decrement.

Suppose that the premiums are payable as the n-year temporary life annuity by (30). Then after the set of commands up to p_2, we add the following commands to obtain the premiums:

Table 2.5 n-year premium

n	$\bar{a}_{30:\bar{n}|}$	$_nP_{30}^{(1)}$	$_nP_{30}^{(2)}$	$_nP_{30}$
1	0.9742	617.98	51.97	669.95
2	1.8984	317.12	26.67	343.79
3	2.7751	216.94	18.24	235.18
4	3.6066	166.92	14.04	180.96
5	4.3949	136.98	11.52	148.50
6	5.1423	117.07	9.85	126.92
7	5.8506	102.90	8.65	111.55
8	6.5216	92.31	7.76	100.08
9	7.1573	84.11	7.07	91.19
10	7.7591	77.59	6.53	84.11

```
x <- 30;
n <- 1:10
j <- m*a^x;
q <- e^j*gamma(p)*(j^(1-p))*(pgamma(a^n, p, j)
     -pgamma(1, p, j))   #ā_{x:n̄|};
q4 <- (1-q-e^(j-f*n-j*a^n))/f   #ā_{x:n̄|};
pr1 <- 1000*e^(b1*x)*q1/q4   #_nP_x^{(1)};
pr2 <- 1000*q3/q4   #_nP_x^{(2)};
pr3 <- pr1+pr2;
d <- round(data.frame(q4, pr1, pr2, pr3), 4);
d1 <- data.frame(n, d);
d1    #Table 2.5;
```

Table 2.5 reports the premiums, payable as the n-year temporary annuity for the whole life insurance. Here the premiums are also higher as compared to those in Table 2.2.

The following R commands compute the premium for the n-year term insurance:

```
a1 <- 0.0008   #A;
b <- 0.00011   #B;
a <- 1.095   #C;
m <- b/log(a, base=exp(1));
e <- exp(1);
del <- 0.05;
b1 <- 0.02;
b2 <- 0.03;
f <- a1+del ;
p <- (-f/log(a, base=exp(1)))+1   #λ1;
f1 <- a1+del-b1;
p1 <- (-f1/log(a, base=exp(1)))+1   #λ2;
f2 <- a1+del-b2;
```

Table 2.6 Premium for
n-year term insurance

n	$\bar{a}_{30:\bar{n}|}$	$P^{1(1)}_{30:\bar{n}|}$	$P^{1(2)}_{30:\bar{n}|}$	$P^{1}_{30:\bar{n}|}$
1	0.9742	3.22	2.00	5.22
2	1.8984	3.41	2.03	5.44
3	2.7751	3.60	2.06	5.66
4	3.6066	3.81	2.09	5.89
5	4.3949	4.03	2.12	6.14
6	5.1423	4.26	2.15	6.40
7	5.8506	4.50	2.18	6.68
8	6.5216	4.76	2.21	6.96
9	7.1573	5.03	2.23	7.26
10	7.7591	5.32	2.26	7.58

```
p2 <- (-f2/log(a, base=exp(1)))+1   #λ3;
x <- 30;
j <- m*a^x   #αx;
n <- 1:10;
q <- e^j*gamma(p)*(j^(1-p))*(pgamma(a^n, p, j)
     -pgamma(1, p, j));
q1 <- e^(j+b1*x)*gamma(p1)*(j^(1-p1))*(pgamma(a^n, p1, j)
      -pgamma(1, p1, j))   #first term in ĀT;
q2 <- e^j*gamma(p2)*(j^(1-p2))*(pgamma(a^n, p2, j)
      -pgamma(1, p2, j));
q3 <- e^(b2*x)*a1*(1-q2-e^(j-f2*n-j*a^n))/f2;
                           #second term in ĀT;
q4 <- (1-q-e^(j-f*n-j*a^n))/f   #āx:n|;
pr1 <- 1000* q1/q4;
pr2 <- 1000*q3/q4;
pr3 <- pr1+pr2;
d <- round(data.frame(q4, pr1, pr2, pr3), 2);
d1 <- data.frame(n, d);
d1    #Table 2.6;
```

The premiums for the *n*-year term insurance are given in Table 2.6.

Next two examples illustrate computation of discrete premiums.

Example 2.3.4 A multiple decrement model with two causes of decrement is given below in terms of the forces of decrement as

$$\mu^{(1)}_{x+t} = BC^{x+t}, \quad t \geq 0, \qquad \mu^{(2)}_{x+t} = A, \quad t \geq 0,\ A \geq 0,\ B \geq 0,\ C \geq 1.$$

Suppose $A = 0.0008$, $B = 0.00011$, and $C = 1.095$. Further, the force of interest is $\delta = 0.05$. The benefit to be payable at the end of year of death is specified as

$1000e^{0.02(x+k+1)}$ units if death is due to cause 1 and $1000e^{0.03(x+k+1)}$ units if the death is due to cause 2.

(i) Find the premium payable as the discrete life annuity due by (30) for the whole life insurance. Decompose the total premium according to two causes of decrement.

(ii) Find the premium payable as the discrete n-year temporary life annuity due by (30) for the whole life insurance, for $n = 1, 2, \ldots, 10$. Decompose the total premium according to two causes of decrement.

(iii) Find the premium payable as the discrete n-year temporary life annuity due by (30), for the n-year term insurance, for $n = 1, 2, \ldots, 10$.

Solution To find the premiums, we have to first find out the actuarial present value of the benefits and the actuarial present value of the annuities corresponding to two modes of premium payments. The actuarial present value of the benefits in the whole life insurance is given by

$$AW_x = \sum_{k=0}^{\infty} v^{k+1} e^{0.02(x+k+1)} P[K = k, J = 1]$$

$$+ \sum_{k=0}^{\infty} v^{k+1} e^{0.03(x+k+1)} P[K = k, J = 2].$$

Similarly, the actuarial present value of the benefits in the n-year term insurance is given by

$$AT^1_{x:\overline{n}|} = \sum_{k=0}^{n-1} v^{k+1} e^{0.02(x+k+1)} P[K = k, J = 1]$$

$$+ \sum_{k=0}^{n-1} v^{k+1} e^{0.03(x+k+1)} P[K = k, J = 2].$$

Using the joint distribution of $K(30)$ and $J(30)$ derived in Example 1.2.5, we can obtain these actuarial present values. Further, the actuarial present value of the whole life annuity and n-year temporary life annuity is given by

$$\ddot{a}_{x:\overline{n}|} = \sum_{k=0}^{n-1} v^k {}_k p_x^{(\tau)} \quad \text{and} \quad \ddot{a}_x = \sum_{k=0}^{\infty} v^k {}_k p_x^{(\tau)},$$

where ${}_k p_x^{(\tau)}$ for the given two decrement model is ${}_k p_x^{(\tau)} = \exp[-Ak - \alpha_x C^k + \alpha_x]$, as derived in Example 1.2.5. The following set of R commands computes all these actuarial present values and the premiums for the two insurance products. Commands at the beginning compute the joint distribution of $K(30)$ and $J(30)$ and are the same as given in Example 1.2.5.

```
a1 <- 0.0008    # A;
b <- 0.00011    # B,
a <- 1.095      # C;
m <- b/log(a, base=exp(1));
e <- exp(1);
f <- (-a1/log(a, base=exp(1)))+1
                #parameter λ as defined in Example 1.2.2;
x <- 30;
k <- 0:69;
j <- m*a^x;
j1 <- m*a^(x+k);
p <- e^(-a1*k+j-j*a^k)  #vector of kp_{30}^{(τ)} for k=0 to 69;
q1 <- e^j1*gamma(f)*(j1^(1-f))*(pgamma(a, f, j1)
        -pgamma(1, f, j1))  #vector of q_{30+k}^{(1)} for k=0 to 69;
q2 <- 1-e^(-a1-j1*a+j1)-q1
                    #vector of q_{30+k}^{(2)} for k=0 to 69;
p1 <- p*q1  #vector of P[K(30)=k, J(30)=1] for k=0 to 69;
p2 <- p*q2  #vector of P[K(30)=k, J(30)=2] for k=0 to 69;
del <- 0.05;
v <- e^(-del);
b1 <- 0.02;
b2 <- 0.03;
x <- 30;
w130 <- e^(b1*x)*sum(p1*(v*e^b1)^(k+1))
                                #first term in A;
w230 <- e^(b2*x)*sum(p2*(v*e^b2)^(k+1))
                                #second term in A;
w30 <- w130+w230  # A;
w130;  w230;  w30;
wa <- e^j*sum(v^k*e^(-a1*k-j*a^k))  #ä_x;
pw1 <- 1000*w130/wa  #premium corresponding to first
                    #cause in whole life insurance;
pw2 <- 1000*w230/wa  #premium corresponding to second
                    #cause in whole life insurance;
pw <- pw1+pw2  #premium corresponding to whole life
                #insurance;
pw1;  pw2;  pw;
ta <- e^j*cumsum(v^k*e^(-a1*k-j*a^k));
nta <- ta[1:10]  #ä_{x:n̄|} for n=1,2,...,10;
pnw1 <- 1000*w130/nta  #n-year premium corresponding to
                    #first cause in whole life insurance;
pnw2 <- 1000*w230/nta  #n-year premium corresponding
                    #to second cause in whole life insurance;
pnw <- pnw1+pnw2  #n-year premium corresponding to whole
                #life insurance;
```

Table 2.7 *n*-year discrete premiums for whole life insurance

n	$_nP^{(1)}$	$_nP^{(2)}$	$_nP$
1	593.07	50.12	643.20
2	304.33	25.72	330.05
3	208.18	17.59	225.77
4	160.18	13.54	173.72
5	131.44	11.11	142.55
6	112.33	9.49	121.83
7	98.73	8.34	107.07
8	88.56	7.49	96.05
9	80.69	6.82	87.51
10	74.43	6.29	80.72

```
d <- round(data.frame(pnw1, pnw2, pnw), 2);
n <- 1:10;
d1 <- data.frame(n, d);
d1   #Table 2.7;
t130 <- e^(b1*x)*cumsum(p1*(v*e^b1)^(k+1))
                                #first term in AT^1_{x:n|};
t230 <- e^(b2*x)*cumsum(p2*(v*e^b2)^(k+1))
                                #second term in AT^1_{x:n|};
t1 <- t130[1:10]   #AT^1_{x:n|} for n = 1,2,...,10, due to cause 1;
t2 <- t230[1:10]   #AT^1_{x:n|} for n = 1,2,...,10, due to cause 2;
pt1 <- 1000*t1/nta  #premium corresponding to first
                    #cause in n-year term insurance;
pt2 <- 1000*t2/nta  #premium corresponding to second
                    #cause in n-year term insurance;
pt <- pt1+pt2  #premium for n-year term insurance;
d3 <- round(data.frame(pt1, pt2, pt), 2);
d4 <- data.frame(n, d3);
d4   #Table 2.8;
```

The first term in AW_x is 0.5931, while the second term is 0.0501, and A is 0.6432. It is to be noted that these values are close and slightly smaller than the corresponding quantities in $\bar{A}$ computed in Example 2.3.2. The premiums corresponding to the first cause, the second cause and total premium in the whole life insurance, payable as the whole life annuity due, are, 35.50, 3.00, and 38.50, respectively. Table 2.7 reports the premiums for the whole life insurance when they are paid as the *n*-year temporary life annuity.

It is to be noted that these premiums are slightly lower than the corresponding continuous premiums. Table 2.8 gives the premiums for the *n*-year term insurance, with split corresponding to two causes of decrement.

Table 2.8 Discrete premium for n-year term insurance

n	$_nPT^{(1)}$	$_nPT^{(2)}$	$_nPT$
1	3.10	1.93	5.02
2	3.27	1.95	5.23
3	3.46	1.98	5.44
4	3.66	2.01	5.67
5	3.86	2.04	5.90
6	4.08	2.07	6.15
7	4.32	2.10	6.42
8	4.56	2.13	6.69
9	4.83	2.15	6.98
10	5.10	2.18	7.28

Here also, these premiums are again close and slightly lower than the corresponding continuous premiums, as displayed in Table 2.6.

In the previous examples we have obtained the premiums corresponding to two causes of decrement and also the total premium. Example 2.3.5 illustrates similar computations for the whole life insurance in the presence of rider.

Example 2.3.5 A multiple decrement model with two causes of decrement is given below in terms of the forces of decrement as

$$\mu_{x+t}^{(1)} = BC^{x+t}, \quad t \geq 0, \qquad \mu_{x+t}^{(2)} = A, \quad t \geq 0, \ A \geq 0, \ B \geq 0, \ C \geq 1.$$

Suppose $A = 0.0008$, $B = 0.00011$, and $C = 1.095$. Further, the force of interest is $\delta = 0.05$. The benefit to be payable at the end of year of death is specified as 1000 units in the whole life insurance contract issued to (30). Extra benefit of 1000 units is payable at the end of year of death if death is due to accident before (30) attains age 65. Find the premium payable as the discrete whole life annuity due by (30) for the whole life insurance. Find the extra premium to be payable as the 35-year temporary life annuity due if death occurs before age 65 due to accident.

Solution Let $J = 1$ if death is nonaccidental and $J = 2$ if death is due to accident. Thus the premium for the base policy is given by

$$P_{30}^{(\tau)} = \frac{AW_{30}}{\ddot{a}_x} = \frac{1000 \sum_{k=0}^{\infty} v^{k+1} P[K=k]}{\sum_{k=0}^{\infty} v^k {}_k p_x^{(\tau)}} = \frac{185.13}{16.71} = 11.08.$$

The extra premium is due to cause 2. So it is given by

$$P_{30}^{(2)} = \frac{AT_{30}}{\ddot{a}_{30:\overline{35}|}} = \frac{1000 \sum_{k=0}^{34} v^{k+1} P[K=k, J=2]}{\sum_{k=0}^{34} v^k {}_k p_x^{(\tau)}} = \frac{11.97}{15.79} = 0.76.$$

These computations are done using the following set of R commands:

```
a1 <- 0.0008   # A;
b <- 0.00011   # B,
a <- 1.095   # C;
m <- b/log(a, base=exp(1));
e <- exp(1);
f <- (-a1/log(a, base=exp(1)))+1 #parameter λ as
                                 #defined in Example 1.2.2;
x <- 30;
k <- 0:69;
j <- m*a^x;
j1 <- m*a^(x+k);
p <- e^(-a1*k +j -j*a^k)   #vector of ₖp₃₀⁽ᵀ⁾ for k=0 to 69;
q1 <- e^j1*gamma(f)*(j1^(1-f))*(pgamma(a, f, j1)
         -pgamma(1, f, j1)) #vector of q₃₀₊ₖ⁽¹⁾ for k=0 to 69;
q2 <- 1-e^(-a1-j1*a+j1)-q1 #vector of q₃₀₊ₖ⁽²⁾ for k=0 to 69;
p1 <- p*q1   #vector of P[K(30)=k,J(30)=1] for k=0 to 69;
p2 <- p*q2   #vector of P[K(30)=k,J(30)=2] for k=0 to 69;
p3 <- p1+p2;
del <- 0.05;
v <- e^(-del);
w130 <- sum(p3*v^(k+1))      # AW₃₀;
wa <- e^j*sum(v^k*e^(-a1*k-j*a^k))   # äₓ;
pw1 <- 1000*w130/wa    #premium for base policy;
w130;  wa;  pw1;
t130 <- cumsum(p2*v^(k+1));
t1 <- t130[35]   # AT₃₀;
ta <- e^j*cumsum(v^k*e^(-a1*k-j*a^k));
nta <- ta[35]   # äₓ:₃₅|;
pt2 <- 1000*t1/nta   #premium for extra benefit;
t1;  nta;  pt2;
```

With the individual life insurance, sometimes there is a rider for disability benefits. For example, during the period of disability, the premiums for the insurance may be waived, or the policy may contain a provision for monthly income for the period of disability. The actuarial present value of the such benefit and hence the premiums can be determined using the multiple decrement model.

Apart from the premium calculation, another important computation in insurance industry is the reserve computation. We now proceed to discuss reserve calculations for the multiple decrement model in the next section.

2.4 Computation of Reserves

Reserve computations for various insurance products, in continuous and discrete setup, are thoroughly discussed in the literature on Actuarial Statistics. In this section we discuss how the reserve computation for single decrement model gets extended to the multiple decrement model. In insurance industry, usually the reserve for the base policy is computed, and a separate reserve is held for the extra benefit. Hence, we discuss reserve calculations for policies with riders, that is, when the underlying survivorship model is a multiple decrement model. The reserve for the base policy can be computed exactly on similar lines as that for single decrement model with the only modification that μ_x, q_x, and p_x will be replaced by $\mu_x^{(\tau)}$, $q_x^{(\tau)}$, and $p_x^{(\tau)}$, respectively. The reserve for the extra benefit can also be computed exactly on similar lines as that for single decrement model, replacing μ_x, q_x, and p_x by $\mu_x^{(j)}$, $q_x^{(j)}$, and $p_x^{(j)}$, respectively, for $j = 1, 2, \ldots, m$. Thus the theoretical development of the formulae for reserves remains the same as that for the single decrement model. The following example illustrates the computational procedure for discrete reserve.

Example 2.4.1 A multiple decrement model with two causes of decrement is given below in terms of the forces of decrement as

$$\mu_{x+t}^{(1)} = BC^{x+t}, \qquad \mu_{x+t}^{(2)} = A, \qquad A \geq 0, \ B \geq 0, \ C \geq 1.$$

Suppose $A = 0.0008$, $B = 0.00011$, and $C = 1.095$. Further, the force of interest is $\delta = 0.05$. The benefit to be payable at the end of year of death is specified as 1000 units in the whole life insurance contract issued to (30). Extra benefit of 1000 units is payable at the end of year of death if death is due to accident before (30) attains age 65. The premium is payable as the discrete whole life annuity due by (30) for base policy and as the 35-year temporary life annuity due for the extra benefit. Find the reserve at $t = 10, 20, 30, 35$ for the base policy and separate reserve for extra benefit.

Solution We have noted that the reserves can be computed exactly on similar lines as that for a single decrement model. Thus the reserve for the base policy and extra benefit is given by the formula

$$_tV(AW_{30}) = AW_{30+t} - P_{30}^{(\tau)}\ddot{a}_{30+t} \quad \text{and}$$

$$_tV(AT_{30}) = AT_{30+t:\overline{35-t|}} - P_{30}^{(2)}\ddot{a}_{30+t:\overline{35-t|}}.$$

In Example 2.3.5 we have calculated $P_{30}^{(\tau)} = 11.08$ and $P_{30}^{(2)} = 0.76$ for the benefit of 1000 units in base policy and extra benefit of 1000 units for accidental death before age 65. The following R commands calculate the actuarial present values required for reserve calculation. In view of having a complete set of commands for reserve, the commands for premium are also included:

```
a1 <- 0.0008    # A;
b <- 0.00011    # B;
a <- 1.095    # C;
m <- b/log(a, base=exp(1));
e <- exp(1);
```

```
f <- (-a1/log(a, base=exp(1)))+1
                #parameter λ as defined in Example 1.2.2;
x <- 30;
k <- 0:69;
j <- m*a^x;
j1 <- m*a^(x+k);
p <- e^(-a1*k+j-j*a^k)   #vector of kp₃₀⁽τ⁾ for k=0 to 69;
q1 <- e^j1*gamma(f)*(j1^(1-f))*(pgamma(a, f, j1)
      -pgamma(1, f, j1))
                        #vector of q⁽¹⁾₃₀₊ₖ for k=0 to 69;
q2 <- 1-e^(-a1-j1*a+j1)-q1
                        #vector of q⁽²⁾₃₀₊ₖ for k=0 to 69;
p1 <- p*q1   #vector of P[K(30)=k,J(30)=1] for k=0 to 69;
p2 <- p*q2   #vector of P[K(30)=k,J(30)=2] for k=0 to 69;
p3 <- p1+p2;
del <- 0.05;
v <- e^(-del);
x <- 30;
w130 <- sum(p3*v^(k+1))   # AW₃₀;
wa <- e^j*sum(v^k*e^(-a1*k-j*a^k))   #ä₃₀;
pw1 <- 1000*w130/wa   #premium for base policy;
t130 <- cumsum(p2*v^(k+1));
t1 <- t130[35]   # AT₃₀:₃₅|;
ta <- e^j*cumsum(v^k*e^(-a1*k-j*a^k));
nta <- ta[35]   #ä₃₀:₃₅|;
pt2 <- 1000*t1/nta   #premium for extra benefit;
x <- 40;
j <- m*a^x;
k <- 0:59;
p <- e^(-a1*k+j-j*a^k);
j1 <- m*a^(x+k);
q1 <- e^j1*gamma(f)*(j1^(1-f))*(pgamma(a, f, j1)
      -pgamma(1, f, j1));
q2 <- 1-e^(-a1-j1*a+j1)-q1;
p1 <- p*q1;
p2 <- p*q2;
p3 <- p1+p2;
w40 <- sum(p3*v^(k+1))   # AW₄₀;
wa40 <- e^j*sum(v^k*e^(-a1*k-j*a^k));
t140 <- cumsum(p2*v^(k+1));
t40 <- t140[25]   # AT₄₀:₂₅|;
ta40 <- e^j*cumsum(v^k*e^(-a1*k-j*a^k));
nta40 <- ta40[25]   #ä₄₀:₂₅|;
x <- 50;
j <- m*a^x;
```

```
k <- 0:49;
p <- e^(-a1*k+j-j*a^k);
j1 <- m*a^(x+k);
q1 <- e^j1*gamma(f)*(j1^(1-f))*(pgamma(a, f, j1)
      -pgamma(1, f, j1));
q2 <- 1-e^(-a1-j1*a+j1)-q1;
p1 <- p*q1;
p2 <- p*q2;
p3 <- p1+p2;
w50 <- sum(p3*v^(k+1))      # AW50;
wa50 <- e^j*sum(v^k*e^(-a1*k-j*a^k));
t150 <- cumsum(p2*v^(k+1));
t50 <- t150[15]    # AT50:15|;
ta50 <- e^j*cumsum(v^k*e^(-a1*k-j*a^k));
nta50 <- ta50[15]    # ä50:15|;
x <- 60;
j <- m*a^x;
k <- 0:39;
p <- e^(-a1*k+j-j*a^k);
j1 <- m*a^(x+k);
q1 <- e^j1*gamma(f)*(j1^(1-f))*(pgamma(a, f, j1)
      -pgamma(1, f, j1));
q2 <- 1-e^(-a1-j1*a+j1)-q1;
p1 <- p*q1;
p2 <- p*q2;
p3 <- p1+p2;
w60 <- sum(p3*v^(k+1))      # AW60;
wa60 <- e^j*sum(v^k*e^(-a1*k-j*a^k));
t160 <- cumsum(p2*v^(k+1));
t60 <- t160[5];
ta60 <- e^j*cumsum(v^k*e^(-a1*k-j*a^k));
nta60 <- ta60[5]    # ä60:5|;
x <- 65;
j <- m*a^x;
k <- 0:34;
p <- e^(-a1*k+j-j*a^k);
j1 <- m*a^(x+k);
q1 <- e^j1*gamma(f)*(j1^(1-f))*(pgamma(a, f, j1)
      -pgamma(1, f, j1));
q2 <- e^(-a1-j1*a+j1)-q1;
p1 <- p*q1;
p2 <- p*q2;
p3 <- p1+p2;
w65 <- sum(p3*v^(k+1))      # AW65;
wa65 <- e^j*sum(v^k*e^(-a1*k-j*a^k))    # ä65;
```

Table 2.9 Discrete reserves for a policy with rider

t	$_tV_{30}^{(\tau)}$	$_tV_{30}^{(2)}$	$_tV_{30}$
10	106.58	−0.02	106.56
20	248.01	−0.04	247.97
30	417.04	−0.04	417.00
35	505.70	0.00	505.70

```
w <- c(w40, w50, w60, w65);
wa <- c(wa40, wa50, wa60, wa65);
vb <- round(1000*(w-pw1*wa), 2)   # tV(AW30);
te <- c(t40, t50, t60, 0);
ta <- c(nta40, nta50, nta60, 0);
vt <- round(1000*(te-pt2*ta), 2)  # tV(AT30);
v1 <- vb+vt;
t <- c(10, 20, 30, 35);
d <- data.frame(t, vb, vt, v1);
d   #Table 2.9;
```

Table 2.9 gives the reserves for base policy and for rider.

The reserve corresponding to extra benefit is negative, implying that the insurer does not have positive liability corresponding to the extra benefit. The negative value may be in view of the fact that the chance of claim due to accidental death is very less, and thus the actuarial present value of the outflow of the company is smaller than that of inflow. This picture is of course for the given model with given set of parameters. We will get different results for different models.

Key Terms   Actuarial present value, Premium, Reserves, Term insurance, Whole life insurance.

2.5 Exercises

2.1 For the 20-year term insurance issued to (30), the following information is given.
   (i) $\mu_{30+t}^{(1)} = 0.0005t$, where (1) represents death by accidental means.
   (ii) $\mu_{30+t}^{(2)} = 0.0025t$, where (2) represents death by other means.
   (iii) The benefit is 2000 units if death occurs by accidental means and 1000 units if death occurs by other means.
   (iv) Benefits are payable at the moment of death.
   Taking $\delta = 0.06$, find the premium payable as (i) the 20-year temporary continuous life annuity, (ii) 20-year temporary life annuity due, (iii) 10-year temporary continuous life annuity, and (iv) 10 year temporary life annuity due.

2.2 A multiple decrement model with 2 causes of decrement is given below in terms
of the forces of decrement as

$$\mu_{x+t}^{(1)} = B_1 C_1^{x+t}, \qquad \mu_{x+t}^{(2)} = B_2 C_2^{x+t}, \qquad B_i \geq 0, \ C_i \geq 1, \ i = 1, 2.$$

Suppose $B_1 = 0.00012$, $C_1 = 1.094$, $B_2 = 0.00014$, and $C_2 = 1.091$. Further,
the force of interest is $\delta = 0.06$. The benefit to be payable at the moment of
death is specified as 1000 units in the whole life insurance contract issued to
(25). Extra benefit of 1000 units is payable at the moment of death if death is
due to accident before (25) attains age 60.

(i) Find the premium payable as the whole life continuous annuity for the
whole life insurance issued to (25). Find the extra premium to be payable
as the 35-year temporary continuous life annuity if death occurs before age
60 due to accident.

(ii) Find the reserve at $t = 10, 20, 35$ for the base policy and separate reserve
for extra benefit.

(iii) Find the premium payable as the discrete whole life annuity due by (25)
for the whole life insurance. Find the extra premium to be payable as the
35-year temporary life annuity due if death occurs before age 60 due to
accident.

(iv) Find the reserve at $t = 10, 20, 35$ for the base policy and separate reserve
for extra benefit.

2.3 A multiple decrement model with two causes of decrement is given below in
terms of the forces of decrement as

$$\mu_{x+t}^{(1)} = B_1 C_1^{x+t}, \qquad \mu_{x+t}^{(2)} = B_2 C_2^{x+t}, \qquad B_i \geq 0, \ C_i \geq 1, \ i = 1, 2.$$

Suppose $B_1 = 0.00012$ and $C_1 = 1.094$, $B_2 = 0.00014$, and $C_2 = 1.091$. Further, the force of interest is $\delta = 0.06$. The benefit to be payable at the end of year
of death is specified as 1000 units in the whole life insurance contract issued to
(25). Extra benefit of 1000 units is payable at the end of year of death if death
is due to accident before (25) attains age 60.

(i) Find the premium payable as the discrete whole life annuity due by (25)
for the whole life insurance. Find the extra premium to be payable as the
35-year temporary life annuity due, if death occurs before age 60 due to
accident.

(ii) Find the reserve at $t = 10, 20, 35$ for the base policy and separate reserve
for extra benefit.

Chapter 3
Defined Benefit Pension Plan

3.1 Introduction

Individuals constantly seek means to enhance their economic security. One cause of economic insecurity is the probable reduction of an individual's earning power at an advanced age. The problem of superannuation exists everywhere. The risk is met through either personal savings, private pensions, or government sponsored programs. For a large number of employers, the formal pension plan approach has proved to be a good solution. In general, a pension is an arrangement to provide people with a regular income after superannuation. Pension plans may be set up by employers, insurance companies, the government, or other institutions such as employer associations or trade unions. Retirement pensions are typically in the form of a guaranteed life annuity, thus insuring against the risk of longevity. Many insurance companies offer pension plans. Insurance companies in USA entered the pension business in 1921, when the Metropolitan Life Insurance Company issued the first group annuity contract.

Pension plan analysis is one of the most active areas for both research and applications. A pension plan is a financial contract where the main pension benefit is in the form of a deferred life annuity. Pension plans are typically employer sponsored plans. Employers sponsor plans for number of reasons. Some of these are as follows: (1) competition for new employees, (2) to facilitate turnover of older employees by ensuring that they can afford to retire, (3) to provide incentive for employees to stay with the employer, (4) pressure from trade unions, (5) to provide tax-efficient method of remunerating employees, and (6) responsibility to employees who have contributed to the success of the company. The pension plan design depends on which of these motivations is the most important to the sponsor.

Pension plans are classified as defined benefit or defined contribution pension plans. A defined benefit pension plan guarantees a certain payout at retirement, according to a fixed formula which usually depends on the member's salary and the number of years of membership in the plan, that is, the number of years of service. For example, suppose that an employee reaches retirement age with n years of service (that is, membership of the pension plan) and pensionable salary S. Then the annual initial pension at retirement is $Sn\alpha$, where α is called the accrual rate and is usually 1–2 %. The pensionable salary may be the average salary of last few months before retirement or may be the average of the salary throughout the employment. These two cases are referred to as the final salary plan and career average salary plan. In India, according to the Maharashtra state Goverment's rule, the pensionable salary is the last basic salary or average of basic salary of last 10 months, whichever is maximum. The monthly pension is 50 % of the pensionable salary plus the allowances. Before February 2009, the monthly pension was defined by the formula

$$(\text{Pensionable salary}/2) \times (\text{total service}/66).$$

The accrual rate is set by consideration of appropriate replacement ratio. The pension plan replacement ratio R is defined as

$$R = \frac{\text{Pension income in the year after retirement}}{\text{Salary in the year before retirement}},$$

where it is assumed that the member survives the year following retirement. The value of the ratio is defined to allow retirees to maintain their preretirement lifestyle. Employer sponsored plans often target 50–70 % as the replacement ratio for an employee with a full career in the company.

A defined contribution plan provides a payout at retirement that is dependent upon the amount of money contributed and the performance of the investment vehicles utilized. Thus, defined contribution pensions work more like a bank account. The employee and employer pay a predetermined contribution, usually a fixed percentage of the salary, into a fund, and the fund earns interest. When the employee leaves or retires, the proceeds are available to provide income throughout retirement. In UK, proceeds are converted to annuity; in USA and Canada there are options: the pensioner may draw funds to live on without necessarily purchasing an annuity from the insurance company.

Some types of retirement plans, such as cash balance plans, combine features of both defined benefit and defined contribution plans. They are often referred to as hybrid plans. Such plan designs have become increasingly popular in the US since 1990. Examples include Cash Balance and Pension Equity plans. Many countries have created funds for their citizens and residents to provide income when they retire (or in some cases become disabled), for example, Employee's Provident Fund Organization in India, National Insurance in UK, or Social Security in the USA. Typically this requires payments throughout the citizen's working life in order to qualify for benefits later on.

Thus, a pension plan in its simplest form is a promise by the employer to pay a periodic benefit, usually for life, to employees who meet the requirements asserted in the plan. As an illustration, suppose that a defined benefit pension plan provides employees with a retirement benefit, usually monthly life annuity of certain amount, after retirement age of 60 or 65, provided that the employee has completed minimum of 5 or 10 years of service with the employer. For an employer, it is of interest to know the cost of this plan at certain time point for all the individuals who retire over the period. The knowledge of the cost will enable him to decide the amount of fund needed to meet the liability of providing pension benefits and thereby help to take the decisions to raise the fund. Thus, the employer has to measure and value at a regular point in time, may be each year, an ever changing group of participants involving active, retired, terminated, and disabled participants. Such a dynamic group is an important aspect of deciding the cost of pension plan, and in this chapter we elaborate on this point.

It is interesting to note that not all the current employees will be entitled to the benefit under the plan. Some may die in service, some may quit the job, and some may become disabled prior to retirement age. Pension plan also specifies the benefit if the individual dies when in active service or becomes disabled and hence has to retire or if he withdraws after some period of service. Thus the cause of termination from active service is a random variable. Further, the future life of retired employees after retirement is also a random variable. For the specified pension benefit, the amount of annual benefit payments under the plan depends upon the number of retired workers. This number, in turn depends on the rate at which already retired individuals die and the rate at which new employees are added to the retirement roles. Similarly the future life of individuals terminating due to other causes also brings in randomness. Employer has to determine the cost of the pension plan for such a group of employees and has to decide the pattern of annual contribution to meet the payments of benefit spelled out in the plan. The only way to determine the true cost of a pension plan would be to wait until the last retired employee has died and then add up all the benefit payments and administrative expenses since the inception of the plan and subtract the investment earnings. However it is totally impractical. Thus, the cost of any pension plan is uncertain until the plan terminates at some date in distant future. As a consequence, it is essential to estimate the ultimate cost of the pension plan with reasonable accuracy and thus arrive at a level of estimated plan contributions. The rate of investment income on the fund generated via contributions is also uncertain. In view of such a variety of uncertainties, statistical tools are essential in pension planning.

In summary, in the estimation of pension cost, it is necessary to make a number of assumptions regarding the factors that affect plan's cost. The first step in the estimation of pension cost is the estimate of benefits paid, expenses, and the investment return. The estimate of benefits paid depends on the benefit provision of the plan, the characteristics such as, age, sex, salary, length of service of the participants of the plan, and the mortality pattern, that is, assumptions regarding the longevity of the participants. Once an estimate of the ultimate cost of the plan is determined, the next step is to determine the contribution required to pay for the estimated cost. Cost methods determine the incidence, that is, occurrence or frequency, and amount of pension costs. An orderly method of finding the cost of the pension plan, in the presence of such multifaceted uncertainty, is known as the funding method. Different sets of assumptions lead to different funding methods. Payments to meet the cost of benefits in pension plan are known as contributions and not premiums as in insurance policies. Of course, the principle to determine amount of contribution remains the same as balancing the outflow of the employer to the inflow via contribution, which may be partly by employer and partly by the employee.

The pensions actuary working with defined benefit pensions must determine appropriate contribution rates to meet the benefits promised, using models suitable to the working patterns of the employees. Sometimes, the employer may want to change the benefit structure, and the actuary is then responsible for assessing the cost and impact of changes. When one company with pension plan takes over another, the actuary must assist with determining the best way to allocate the assets from the two plans and how to merge the benefits.

In this chapter we discuss the valuation of pension plans, and the next chapter discusses various methods of funding the pension costs. In the next section, we discuss how to compute the actuarial present value of the benefit under a pension plan.

3.2 Actuarial Present Value of Pension Benefit

The ultimate cost of the plan is defined as the benefits paid and administrative expenses less the investment earnings. In this chapter we will concentrate on the first component, that is, benefits paid. The amount of benefits paid depends upon the number of workers who will ultimately be entitled to receive benefits under the plan. This number depends on following four factors: (i) mortality rates among active employees, (ii) rates of disabilities among active employees under a plan that offers disability benefit, (iii) rate of withdrawals from employment, and (iv) rates of retirement at different eligible ages. Further, the amount of benefit paid also depends on the length of benefit period, which in turn depends on the longevity of retired members. Therefore assumptions need to be made about the mortality among the retired members. The last factor affecting the total amount paid under the plan is the amount of pension paid to each retired member under the plan. In general the benefit formula includes the salary progression, amount of service, and inflation factor.

In Chap. 1 we studied multiple decrement model when there are multiple modes of exit from a given status of an individual. A major application of multiple decrement models is in valuation of pension plans. Suppose that an individual is a member of a group of employees of an employer. The individual may exit the group by any one of the following modes: retirement at the mandatory age of retirement, death during employment, disability to work or withdrawal from the employment, may be joining another employer. Benefit to the individual changes according to the mode of exit. Benefit on retirement is the monthly life annuity depending on the service period and the skill of the individual. For example, the retirement benefit may be the monthly whole life annuity due or may be the 10-year certain and life annuity, payable monthly. The initial pension may be at an annual rate of 0.02 multiplied by the final 5-year average salary multiplied by the total number of years of service. If death occurs before retirement age, a lump sum may be payable to the beneficiary. The benefit amount may be 3 times or 5 times the salary at the time of death. Sometimes a pension is given to the member's surviving spouse. If an individual withdraws from the employment, then there can be a deferred pension, depending on the years of service of the individual with the employer, or the individual's accumulated contribution may be returned. For example, for an employee with at least 20 years of service before withdrawal, the benefit formula may be the same as that for the employees retiring at mandatory age, but the initial payment of the annuity may be deferred until the earliest eligible age of retirement if the employee would have continued in active status. In case of disability, there may be additional benefit till the individual recovers completely. After recovery there may be life annuity of pension. As a consequence, the actuarial present value of the benefits depends on the cause of exit from the group along with the future life time of an individual. The multiple decrement model thus becomes a foundation of these computations. Contributions to meet the cost of benefits are payable in various portions by the individual and the employer. To determine the rate of contribution in pension funds and to value the pension fund at specific time, it is necessary to find the actuarial present value of the benefits. The theory developed in Chaps. 1 and 2 will be applicable to find the actuarial present value of the benefits and, as a consequence, to decide contributions in pension plans.

Two sets of assumptions are needed to determine the actuarial present value of benefits of various types. One set is labeled as demographic, which includes the first eligible age for retirement, the mandatory age of retirement, and survival functions for retired lives, disabled lives, and lives who have withdrawn. The second set of assumptions is a set of economic assumptions, which includes assumptions about the salary scale function and about investment return or rate of interest.

We begin the discussion with the demographic assumptions. Using the same notation as in Chap. 1, a random variable $J(x) \equiv J$ describes the cause of decrement. In the context of four modes of exit from the active service, we define the random variable J as follows:

$$J = 1 \quad \text{if death in active service,}$$

$$J = 2 \quad \text{if withdrawal from service,}$$

$$J = 3 \quad \text{if disability occurs in active service,}$$

$$J = 4 \quad \text{if retirement due to age-service.}$$

The marginal distribution of J given by $h_j = P[J = j]$ is denoted as follows:

$$h_1 = h_d = P[J = 1], \qquad h_2 = h_w = P[J = 2],$$

$$h_3 = h_i = P[J = 3] \quad \text{and} \quad h_4 = h_r = P[J = 4].$$

Decrement corresponding to $J = 4$ is usually referred to as retirement for age-service. Further, $\mu_x^{(w)}$, $\mu_x^{(d)}$, $\mu_x^{(i)}$, and $\mu_x^{(r)}$ denote the forces of decrement at age x due to withdrawal from service, death in service, retirement for disability, and retirement due to age-service, respectively. Similarly, $q_x^{(w)}$, $q_x^{(d)}$, $q_x^{(i)}$, and $q_x^{(r)}$ denote the probability of decrement in age $(x, x+1)$ due to withdrawal from service, death in service, retirement for disability, and retirement due to age-service respectively.

Let $p_x^{(\tau)}$ denote the probability that individual of age x remains in active service till age $x + 1$. Let $d_x^{(w)}$, $d_x^{(d)}$, $d_x^{(i)}$, and $d_x^{(r)}$ denote the expected number of decrements in age $(x, x + 1)$ due to withdrawal from service, death in service, retirement for disability, and retirement due to age-service, respectively, and let $l_x^{(\tau)}$ denote the number of active members at the beginning of the interval $(x, x + 1)$. As discussed in Chap. 1, $q_x^{(w)}$, $q_x^{(d)}$, $q_x^{(i)}$, and $q_x^{(r)}$ and hence the survival probability can be obtained from the forces of mortality. As derived in Chap. 1, we have the following relation among these functions:

$$l_{x+1}^{(\tau)} = l_x^{(\tau)} p_x^{(\tau)}, \qquad l_x^{(\tau)} q_x^{(w)} = d_x^{(w)},$$

$$l_x^{(\tau)} q_x^{(d)} = d_x^{(d)}, \qquad l_x^{(\tau)} q_x^{(i)} = d_x^{(i)}, \qquad l_x^{(\tau)} q_x^{(r)} = d_x^{(r)}.$$

Further, $_k p_x^{(\tau)} = \frac{l_{x+k}^{(\tau)}}{l_x^{(\tau)}}$ denotes the probability that (x) remains in service for next k years. The following example illustrates the computation of certain probabilities using the forces of decrements.

Example 3.2.1 Suppose that a member of the pension plan, joining the plan at age 25, exits the plan by any one of the four causes of exit: withdrawal, disability retirement, death in service, and age retirement. The multiple decrement model for these four causes of decrement is specified in terms of forces of decrements as follows:

$$\mu_x^{(w)} = \begin{cases} w_1 = 0.13 & \text{if } 25 \le x < 30, \\ w_2 = 0.07 & \text{if } 30 \le x < 40, \\ w_3 = 0.02 & \text{if } 40 \le x < 55, \\ 0 & \text{if } x \ge 55, \end{cases}$$

$$\mu_x^{(r)} = \begin{cases} 0 & \text{if } x < 55, \\ r_1 = 0.06 & \text{if } 55 \le x < 65, \end{cases}$$

$\mu_x^{(i)} = i_1 = 0.005$, and $\mu_x^{(d)} = A + BC^x$, with $A = 0.0007$, $B = 0.0001151$, and $C = 1.096$.

Further, 30 % of the members surviving to age 55 retire at age 55, and 65 is the mandatory age of retirement. Calculate the probability that a member of age 25

(i) retires at exact age 65;
(ii) exits the plan due to (a) withdrawal, (b) age-service retirement, (c) disability, and (d) death in service.

Solution (i) The mandatory age of retirement is 65; hence all those who are in service up to 65 retire at 65. Thus, the probability that a member of age 25 retires at age 65 is the probability that he is in service up to 65 and is given by $_{40}p_{25}^{(\tau)}$. Since there are discontinuities in the withdrawal and retirement forces of decrement, we have to consider separately the periods before and after the points of discontinuities.

Let $m = \frac{B}{\log C}$. For $0 < t < 5$,

$$_t p_{25}^{(\tau)} = \exp\left\{ -\int_0^t \left(\mu_{25+s}^{(w)} + \mu_{25+s}^{(i)} + \mu_{25+s}^{(d)} \right) ds \right\}$$

$$= \exp\{-((A + w_1 + i_1)t + mC^{25}(C^t - 1))\}.$$

With $t = 5$, $_5 p_{25}^{(\tau)} = 0.503727$.
  For $5 \le t < 15$,

$$_t p_{25}^{(\tau)} = {}_5 p_{25}^{(\tau)} \, {}_{t-5} p_{30}^{(\tau)}$$

$$= {}_5 p_{25}^{(\tau)} \exp\left\{ -\int_0^{t-5} \left(\mu_{30+s}^{(w)} + \mu_{30+s}^{(i)} + \mu_{30+s}^{(d)} \right) ds \right\}$$

$$= {}_5 p_{25}^{(\tau)} \exp\{-((A + w_2 + i_1)(t - 5) + mC^{30}(C^{t-5} - 1))\}.$$

Hence,

$$_{15} p_{25}^{(\tau)} = {}_5 p_{25}^{(\tau)} \, {}_{10} p_{30}^{(\tau)} = 0.503727 \times 0.455445 = 0.229420.$$

For $15 \le t < 30$,

$$_t p_{25}^{(\tau)} = {}_5 p_{25}^{(\tau)} \, {}_{10} p_{30}^{(\tau)} \, {}_{t-15} p_{40}^{(\tau)}$$

$$= {}_5 p_{25}^{(\tau)} \, {}_{10} p_{30}^{(\tau)} \exp\left\{ -\int_0^{t-15} \left(\mu_{40+s}^{(w)} + \mu_{40+s}^{(i)} + \mu_{40+s}^{(d)} \right) ds \right\}$$

$$= {}_5 p_{25}^{(\tau)} \, {}_{10} p_{30}^{(\tau)} \exp\{-((A + w_3 + i_1)(t - 15) + mC^{40}(C^{t-15} - 1))\}.$$

Hence,

$$_{30} p_{25}^{(\tau)} = {}_5 p_{25}^{(\tau)} \, {}_{10} p_{30}^{(\tau)} \, {}_{15} p_{40}^{(\tau)} = 0.503727 \times 0.455445 \times 0.588215 = 0.134948.$$

At age 55, 30 % retire so the probability of remaining in the plan, denoted $_{30_+} p_{25}^{(\tau)}$, is given by

$$_{30_+} p_{25}^{(\tau)} = 0.7 \times {}_{30} p_{25}^{(\tau)} = 0.094464.$$

For $30 \le t < 40$,

$$_t p_{25}^{(\tau)} = {}_{30_+} p_{25}^{(\tau)} \exp\left\{ -\int_0^{t-30} \left(\mu_{55+s}^{(r)} + \mu_{55+s}^{(i)} + \mu_{55+s}^{(d)} \right) ds \right\}$$

$$= {}_{30_+} p_{25}^{(\tau)} \exp\{-((A + r_1 + i_1)(t - 30) + mC^{55}(C^{t-30} - 1))\}.$$

Thus the probability of remaining in the plan up to 65, that is, the probability of retirement at exact age 65, denoted $_{40} p_{25}^{(\tau)}$, is given by $_{40} p_{25}^{(\tau)} = 0.094464 \times 0.387277 = 0.036584$.

Thus, the expected number of retirements at 65 is 3658 if we assume that 100000 individuals join the pension plan at age 25 and the group suffers decrements according to specified forces.

(ii) To compute the probability that the member exits the plan due to withdrawal, we note that withdrawals are allowed up to 55. Again the force of withdrawal has discontinuities, so to compute this probability, we split the period appropriately as done in (i). The probability of withdrawal of (25) by age 30, denoted $_5q_{25}^{(w)}$, is given by

$$_5q_{25}^{(w)} = \int_0^5 {_sp_{25}^{(\tau)}} \mu_{25+s}^{(w)} \, ds = w_1 \int_0^5 {_sp_{25}^{(\tau)}}$$

$$= w_1 \int_0^5 \exp\{-((A + w_1 + i_1)s + mC^{25}(C^s - 1))\} \, ds.$$

We compute the value of the integral as a Riemann sum. The interval $(0, 5)$ is partitioned into 100000 intervals to get

$$\int_0^5 {_sp_{25}^{(\tau)}} \, ds = 3.619611.$$

Hence, $_5q_{25}^{(w)} = 0.470549$. The probability of withdrawal of (25) between ages 30 to 40 is given by

$$_5p_{25}^{(\tau)} \, _{10}q_{30}^{(w)} = {_5p_{25}^{(\tau)}} \int_0^{10} {_sp_{30}^{(\tau)}} \mu_{30+s}^{(w)} \, ds$$

$$= {_5p_{25}^{(\tau)}} w_2 \int_0^{10} {_sp_{30}^{(\tau)}} \, ds$$

$$= {_5p_{25}^{(\tau)}} w_2 \int_0^{10} \exp\{-((A + w_2 + i_1)s + mC^{30}(C^s - 1))\} \, ds$$

$$= 0.503727 \times 0.07 \times 6.939017 = 0.244676,$$

where the value of the integral is obtained by numerical integration. The probability of decrement due to withdrawal for (25) between ages 40 to 55 is given by

$$_{15}p_{25}^{(\tau)} \, _{15}q_{40}^{(w)} = {_{15}p_{25}^{(\tau)}} \int_0^{15} {_sp_{40}^{(\tau)}} \mu_{40+s}^{(w)} \, ds$$

$$= {_{15}p_{25}^{(\tau)}} w_3 \int_0^{15} {_sp_{40}^{(\tau)}} \, ds$$

$$= {_{15}p_{25}^{(\tau)}} w_3 \int_0^{15} \exp\{-((A + w_3 + i_1)s + mC^{40}(C^s - 1))\} \, ds$$

$$= 0.229420 \times 0.02 \times 11.8262 = 0.054263,$$

where the value of the integral is again obtained by numerical integration.

Thus the probability of withdrawal of (25) is given by

$$h_w = 0.470549 + 0.244676 + 0.054263 = 0.769488,$$

which is quite high. We compute the probability of decrement due to disability for (25) on similar lines. The probability of disability by age 30, denoted $_5q_{25}^{(i)}$, is given by

$$_5q_{25}^{(i)} = \int_0^5 {}_sp_{25}^{(\tau)} \mu_{25+s}^{(i)} \, ds = i_1 \int_0^5 {}_sp_{25}^{(\tau)} = 0.005 \times 3.619611 = 0.018098.$$

The probability of disability between ages 30 to 40 is given by

$$_5p_{25}^{(\tau)} {}_{10}q_{30}^{(i)} = {}_5p_{25}^{(\tau)} \int_0^{10} {}_sp_{30}^{(\tau)} \mu_{30+s}^{(i)} \, ds$$

$$= {}_5p_{25}^{(\tau)} i_1 \int_0^{10} {}_sp_{30}^{(\tau)} \, ds$$

$$= 0.503727 \times 0.005 \times 6.939017 = 0.017477.$$

The probability of disability between ages 40 to 55 is given by

$$_{15}p_{25}^{(\tau)} {}_{15}q_{40}^{(i)} = {}_{15}p_{25}^{(\tau)} \int_0^{15} {}_sp_{40}^{(\tau)} \mu_{40+s}^{(i)} \, ds$$

$$= {}_{15}p_{25}^{(\tau)} i_1 \int_0^{15} {}_sp_{40}^{(\tau)} \, ds$$

$$= 0.229420 \times 0.005 \times 11.8262 = 0.013566.$$

The probability of disability between ages 55 to 65 is given by

$$_{30+}p_{25}^{(\tau)} {}_{15}q_{55}^{(i)} = {}_{30+}p_{25}^{(\tau)} \int_0^{10} {}_sp_{55}^{(\tau)} \mu_{55+s}^{(i)} \, ds$$

$$= {}_{30+}p_{25}^{(\tau)} i_1 \int_0^{10} {}_sp_{55}^{(\tau)} \, ds$$

$$= {}_{30+}p_{25}^{(\tau)} i_1 \int_0^{10} \exp\{-\left((A + r_1 + i_1)s + mC^{55}(C^s - 1)\right)\} \, ds$$

$$= 0.094464 \times 0.005 \times 6.598791 = 0.003117.$$

Thus the probability of decrement due to disability for (25) is given by

$$h_i = 0.018098 + 0.017477 + 0.013566 + 0.003117 = 0.052257.$$

The probability of decrement due to age-service retirement is the sum of the probabilities of exact age retirement at 55 and 65 and the probability due to retirement between 55 and 65. The probability of decrement due to age-service retirement at exact age 55 is

$$0.3 \times {}_{30}p_{25}^{(\tau)} = 0.3 \times 0.134948 = 0.040484.$$

The probability of decrement due to age-service retirement at exact age 65 is computed as 0.036584. The probability of retirement between ages 55 and 65 is given by

$$\begin{aligned}
{}_{30+}p_{25}^{(\tau)} \, {}_{10}q_{55}^{(r)} &= {}_{30+}p_{25}^{(\tau)} \int_0^{10} {}_s p_{55}^{(\tau)} \mu_{55+s}^{(r)} \, ds \\
&= {}_{30+}p_{25}^{(\tau)} r_1 \int_0^{10} {}_s p_{55}^{(\tau)} \, ds \\
&= 0.094464 \times 0.06 \times 6.598791 = 0.037401.
\end{aligned}$$

Thus the probability of decrement due to age-service retirement for (25) is given by

$$h_r = 0.040484 + 0.036584 + 0.037401 = 0.114469.$$

The last step is to find the probability of decrement due to death in active service. We have to calculate this probability again for different periods as the survival probability differs from period to period. The probability for (25) of death by age 30, denoted ${}_5q_{25}^{(d)}$, is given by

$$\begin{aligned}
{}_5q_{25}^{(d)} &= \int_0^5 {}_s p_{25}^{(\tau)} \mu_{25+s}^{(d)} \, ds = \int_0^5 {}_s p_{25}^{(\tau)} \left(A + BC^{25+s} \right) ds \\
&= A \int_0^5 {}_s p_{25}^{(\tau)} \, ds + BC^{25} \int_0^5 {}_s p_{25}^{(\tau)} C^s \, ds = 0.007627.
\end{aligned}$$

Here also we find the values of the integrals by numerical integration. The probability of death between ages 30 to 40 is given by

$$\begin{aligned}
{}_5p_{25}^{(\tau)} \, {}_{10}q_{30}^{(d)} &= {}_5p_{25}^{(\tau)} \int_0^{10} {}_s p_{30}^{(\tau)} \mu_{30+s}^{(d)} \, ds \\
&= {}_5p_{25}^{(\tau)} \left[A \int_0^{10} {}_s p_{30}^{(\tau)} \, ds + BC^{30} \int_0^{10} {}_s p_{30}^{(\tau)} C^s \, ds \right] = 0.012155.
\end{aligned}$$

The probability of death between ages 40 to 55 is given by

$${}_{15}p_{25}^{(\tau)} \, {}_{15}q_{40}^{(d)} = {}_{15}p_{25}^{(\tau)} \int_0^{15} {}_s p_{40}^{(\tau)} \mu_{40+s}^{(d)} \, ds = 0.026643$$

The probability of death between ages 55 to 65 is given by

$${}_{30+}p_{25}^{(\tau)} \, {}_{15}q_{55}^{(d)} = {}_{30+}p_{25}^{(\tau)} \int_0^{10} {}_s p_{55}^{(\tau)} \mu_{55+s}^{(d)} \, ds = 0.017363.$$

Thus the probability of decrement due to death for (25) is given by

$$h_d = 0.007627 + 0.012155 + 0.026643 + 0.017363 = 0.063792.$$

It is to be noted that $h_w + h_i + h_r + h_d = 1$ as it should be.

The following is an R code for these computations:

```
a <- 0.0007; b <- 0.0001151; c1 <- 1.096; m <- b/log(c1);
w1 <- 0.13;   w2 <- 0.07;   w3 <- 0.02;
r1 <- 0.06;   i1 <- 0.005;
t1 <- 5;
p1 <- exp(-((a+w1+i1)*t1+m*c1^25*(c1^t1-1)));   p1;
ps1 <- p1;   ps1   # 5p(τ)25 ;
t2 <- 10;
p2 <- exp(-((a+w2+i1)*t2+m*c1^30*(c1^t2-1)));   p2;
ps2 <- p1*p2;   ps2   # 15p(τ)25 ;
t3 <- 15;
p3 <- exp(-((a+w3+i1)*t3+m*c1^40*(c1^t3-1)));   p3;
ps3 <- p1*p2*p3;   ps3   # 30p(τ)25 ;
ps4 <- ps3*0.7;   ps4   # 30+p(τ)25 ;
p4 <- exp(-((a+r1+i1)*t2+m*c1^55*(c1^t2-1)));   p4;
ps5 <- ps4*p4;ps5   # 40p(τ)25 ;
s1 <- seq(0,  5, 0.00001);
s2 <- seq(0, 10, 0.00001);
s3 <- seq(0, 15, 0.00001);
p5 <- sum(exp(-((a+w1+i1)*s1+m*c1^25*(c1^s1-1)))
       *0.00001);   p5;
qw1 <- w1*p5;   qw1   # 5q(w)25 ;
p6 <- sum(exp(-((a+w2+i1)*s2+m*c1^30*(c1^s2-1)))
       *0.00001);   p6   # 10q(w)30 ;
qw2 <- p1*w2*p6;   qw2;
p7 <- sum(exp(-((a+w3+i1)*s3+m*c1^40*(c1^s3-1)))
       *0.00001);   p7   # 15q(w)40 ;
qw3 <- p1*p2*w3*p7;   qw3;
qw <- qw1+qw2+qw3;   qw   #probability of withdrawal hw;
qi1 <- i1*p5;   qi1   # 5q(i)25 ;
qi2 <- p1*i1*p6;   qi2   # 5p(τ)25 10q(i)30 ;
qi3 <- p1*p2*i1*p7;   qi3   # 15p(τ)25 15q(i)40 ;
p8 <- sum(exp(-((a+r1+i1)*s2+m*c1^55*(c1^s2-1)))
       *0.00001);   p8;
qi4 <- p1*p2*p3*0.7*p8*i1;   qi4   # 30+p(τ)25 10q(i)55 ;
qi <- qi1+qi2+qi3+qi4;   qi
                        #probability of disability hi;
qr1 <- p1*p2*p3*0.3;   qr1
            #probability of retirement at exact age 55;
```

```
qr2 <- p1*p2*p3*0.7*p4;    qr2
            #probability of retirement at exact age 65;
qr3 <- p1*p2*p3*0.7*p8*r1;    qr3    # 30+p25(τ) 10q55(r) ;
qr <- qr1+qr2+qr3;    qr  #probability of retirement hr;
p9 <- sum(exp(-((a+w1+i1)*s1+m*c1^(25)*(c1^s1-1)))
        *c1^s1*0.00001);    p9;
qd1 <- a*p5+b*c1^25*p9;    qd1  # 5q25(d) ;
p10 <- sum(exp(-((a+w2+i1)*s2+m*c1^(30)*(c1^s2-1)))
        *c1^s2*0.00001);    p10;
qd2 <- p1*(a*p6+b*c1^30*p10);    qd2  # 5p25(τ) 10q30(d) ;
p11 <- sum(exp(-((a+w3+i1)*s3+m*c1^(40)*(c1^s3-1)))
        *c1^s3*0.00001);    p11;
qd3 <- p1*p2*(a*p7+b*c1^40*p11);    qd3  # 15p25(τ) 15q30(d) ;
p12 <- sum(exp(-((a+r1+i1)*s2+m*c1^(55)*(c1^s2-1)))
        *c1^s2*0.00001);    p12;
qd4 <- p1*p2*p3*0.7*(a*p8+b*c1^55*p12);    qd4 # 30+p25(τ) 10q55(d) ;
qd <- qd1+qd2+qd3+qd4;    qd  #probability of death hd;
q <- qw+qi+qr+qd;    q  #approximately 1;
```

From Example 3.2.1 it is clear that the procedure to compute various probabilities of decrement when the forces of decrement are specified is rather lengthy and tedious as it involves numerical integration. As discussed in Chap. 1, once we have been given the forces of decrement, we can obtain the expressions for survival and decrement probabilities. In the setup of Example 3.2.1 we do not get explicit expressions, but we can find numerical values of these probabilities using numerical integration wherever needed. Once these probabilities are available, we can prepare a multiple decrement table, with some radix, specifying the expected number of survivals and decrements due to four causes. The multiple decrement table is then helpful to compute all the probabilities computed in Example 3.2.1 very easily. Corresponding to given forces of decrement, we first find $p_x^{(\tau)}$ for $x = 25$ to 64, taking into account the fact that withdrawal rates and age service retirement rates change periodically. Once we have the $p_x^{(\tau)}$ values, we get $l_x^{(\tau)}$. Next step is to find the $q_x^{(j)}$ values for four causes of decrement and the corresponding $d_x^{(j)}$ values. The following is a set of R commands to construct a table of survival and decrement probabilities (Table 3.1) and the corresponding table of the expected number of survivals and decrements (Table 3.2) due to four causes, corresponding to the forces of decrements as given in Example 3.2.1:

```
a <- 0.0007    #A;
b <- 0.0001151  #B;
c1 <- 1.096    #C;
m <- b/log(c1);
w1 <- 0.13;    w2 <- 0.07;    w3 <- 0.02 #withdrawal rates;
r1 <- 0.06  #retirement rates;
i1 <- 0.005  #disability rate;
```

```
x1 <- 25:29   #ages from 25 to 29;
p1 <- exp(-((a+w1+i1)+m*c1^x1*(c1-1)));   p1
                                # p_x^{(τ)} for x = 25 to 29;
x2 <- 30:39
p2 <- exp(-((a+w2+i1)+m*c1^x2*(c1-1)));   p2
                                # p_x^{(τ)} for x = 30 to 39;
x3 <- 40:54
p3 <- exp(-((a+w3+i1)+m*c1^x3*(c1-1)));   p3
                                # p_x^{(τ)} for x = 40 to 54;
p55 <- p3[15]*0.7;   p55   # p_{55-}^{(τ)};
x4 <- 55:64
p4 <- exp(-((a+r1+i1)+m*c1^x4*(c1-1)));   p4
                                # p_x^{(τ)} for x = 55 to 64;
xl55 <- "55L"   #lower limit of the age group (55-56);
x <- c(x1, x2, x3, xl55, x4, 65);   x
px <- c(p1, p2, p3, p55, p4, 0);   px  #vector of p_x^{(τ)};
lx1 <- c(100000, 1:30)
for(i in 1:30)
   {
   lx1[i+1] <- lx1[i]*px[i]
   }
lx2 <- 0.7*lx1[31];
lx3 <- c(lx2,1:10)
for(i in 1:10)
   {
   lx3 [i+1] <- lx3[i]*p4[i]
   }
lx <- c(lx1,lx3);   lx   #vector of l_x^{(τ)};
s <- seq(0, 1, 0.00001)
qx1d <- 0
for(i in 1:5)
   {
   qx1d[i] <- sum((a+b*c1^(i+24)*c1^s)*exp(-((a+w1+i1)*s
            +m*c1^(i+24)*(c1^s-1)))*0.00001)
   }
   qx1d
qx2d <- 1:10
for(i in 1:10)
   {
   qx2d[i] <- sum((a+b*c1^(i+29)*c1^s)*exp(-((a+w2+i1)*s
            +m*c1^(i+29)*(c1^s-1)))*0.00001)
   }
   qx2d
qx3d <- 1:15
for(i in 1:15)
```

```
{
qx3d[i] <- sum((a+b*c1^(i+39)*c1^s)*exp(-((a+w3+i1)*s
          +m*c1^(i+39)*(c1^s-1)))*0.00001)
}
qx3d
qx4d <- 1:10
for(i in 1:10)
  {
  qx4d[i] <- sum((a+b*c1^(i+54)*c1^s)*exp(-((a+r1+i1)*s
          +m*c1^(i+54)*(c1^s-1)))*0.00001)
  }
  qx4d
qxd <- c(qx1d, qx2d, qx3d, 0, qx4d, 0)   #vector of qx^(d);
dxd <- lx*qxd;   dxd   #vector of dx^(d);
sum(dxd)/100000
(dxd[1]+ dxd[2]+dxd[3]+dxd[4]+dxd[5])/100000
                            #probability of death hd;
qx1w <- 0
for(i in 1:5)
  {
  qx1w[i] <- w1*sum(exp(-((a+w1+i1)*s+m*c1^(i+24)
          *(c1^s-1)))*0.00001)
  }
  qx1w
qx2w <- 0
for(i in 1:10)
  {
  qx2w[i] <- w2*sum(exp(-((a+w2+i1)*s+m*c1^(i+29)
          *(c1^s-1)))*0.00001)
  }
  qx2w
qx3w <- 0
for(i in 1:15)
  {
  qx3w[i] <- w3*sum(exp(-((a+w3+i1)*s+m*c1^(i+39)
          *(c1^s-1)))*0.00001)
  }
   qx3w
qxw <- c(qx1w, qx2w, qx3w, rep(0, 12))   #vector of qx^(w);
dxw <- lx*qxw;   dxw   #vector of dx^(w);
dw25 <- dxw[1]+ dxw[2]+ dxw[3]+ dxw[4]+ dxw[5];
qw25 <- dw25/100000;  qw25 #probability of withdrawal hw;
sum(dxw)/100000   #hw = qw25;
qx1dis <- 0
for(i in 1:5)
```

```
  qx1dis[i] <- i1*sum(exp(-((a+w1+i1)*s+m*c1^(i+24)
              *(c1^s-1)))*0.00001);   qx1dis
qx2dis <- 0
for(i in 1:10)
  {
  qx2dis[i] <- i1*sum(exp(-((a+w2+i1)*s+m*c1^(i+29)
              *(c1^s-1)))*0.00001)
  }
  qx2dis
qx3dis <- 0
for(i in 1:15)
  {
  qx3dis[i] <- i1*sum(exp(-((a+w3+i1)*s+m*c1^(i+39)
              *(c1^s-1)))*0.00001)
  }
  qx3dis;
qx4dis <- 0
for(i in 1:10)
  {
  qx4dis[i] <- i1*sum(exp(-((a+r1+i1)*s+m*c1^(i+54)
              *(c1^s-1)))*0.00001)
  }
qx4dis;
qxdis <- c(qx1dis, qx2dis, qx3dis, 0, qx4dis ,0)
```
                                                          #vector of $q_x^{(i)}$;
```
dxdis <- lx*qxdis;   dxdis  #vector of
```
$d_x^{(i)}$;
```
sum(dxdis)/100000  #probability of disability
```
h_i;
```
p5 <- c(px[1], 2:30);
for(i in 2:30)
  {
  p5[i] <- px[i]*p5[i-1]
  }
p5
p6 <- p5[30]*0.7;   p6;
p7 <- c(p6, 1:10);
for(i in 1:10)
  {
  p7[i+1] <- p4[i]*p7[i]
  }
p7;
p8 <- c(p5, p7);   p8
k <- c(0:30, x155, 31:40);
kp25 <- c(1, p8)  #vector of
```
$_kp_{25}^{(\tau)}$;
```
d4 <- data.frame(k, kp25);   d4
lx <- 100000*kp25  #vector of
```
$l_x^{(\tau)}$;

```
d5 <- data.frame(x, lx);    d5
q55r <- 0.3*kp25[31];   q55r;
qx4r <- 0
for(i in 1:10)
  {
  qx4r[i] <- r1*sum(exp(-((a+r1+i1)*s+m*c1^(i+54)
           *(c1^s-1)))*0.00001)
  }
qx4r;
q65r <- 1;
qxr <- c(rep(0, 30), q55r, qx4r, q65r)   #vector of q_x^{(r)};
d55r <- 100000*q55r;
dxr1 <- lx[32:41]*qx4r;    dxr1;
d65r <- 100000*kp25[42];
dxr <- c(rep(0, 30), d55r, dxr1, d65r)   #vector of d_x^{(r)};
(sum(dxr1)+d65r+d55r)/100000
                          #probability of retirement h_r;
d1 <- data.frame(x, qxd, qxw, qxdis, qxr, px);   d1
                          # Table 3.1;
dxr <- c(rep(0, 30), d55r, dxr1, d65r)
d2 <- data.frame(x, lx, dxd, dxw, dxdis, dxr);   d2
                          # Table 3.2;
lx1 <- c(100000, 1:41)
for(i in 1:41)
lx1[i+1] <- lx1[i]-(dxd[i]+dxw[i]+dxdis[i]+dxr[i]);   lx1
                          #vector of l_x^{(τ)};
d3 <- data.frame(x, lx1);   d3;
d4 <- data.frame(dxd, dxw, dxdis, dxr);
d5 <- colSums(d4);   d5;
                    #gives the second last row of Table 3.2;
```

The second last row of Table 3.2 presents the expected number of decrements due to four causes. From Table 3.2 we note that the number of members active in service till age 65 is 3658.36. All those who are in service till age 65, retire at 65. Hence, the probability that a member of age 25 retires at exact age 65, as asked in Example 3.2.1(i), can be easily computed as $3658.36/100000 = 0.036584$. Further, from the last row of Table 3.2, the probability that a member of age 25 exits the plan due to withdrawal is obtained as (sum of withdrawals at various ages between 25 to 55)/$100000 = 76949.48/10000 = 0.76949$. Similarly, the probability that a member of age 25 exits the plan due to death is $6378.75/10000 = 0.063787$, the probability that a member of age 25 exits the plan due to disability is $5225.79/100000 = 0.052257$, and the probability that a member of age 25 exits the plan due to age service retirement is $11446.90/100000 = 0.114469$. The last row of Table 3.2 presents these probabilities. It is to be noted that the last row of Table 3.2 displays the probability distribution of $J(25)$.

Table 3.1 Decrement and survival probabilities

Age x	$q_x^{(d)}$	$q_x^{(w)}$	$q_x^{(i)}$	$q_x^{(r)}$	$p_x^{(\tau)}$
25	0.001767	0.121497	0.004673	0.000000	0.872064
26	0.001874	0.121490	0.004673	0.000000	0.871964
27	0.001991	0.121483	0.004672	0.000000	0.871855
28	0.002119	0.121475	0.004672	0.000000	0.871735
29	0.002260	0.121466	0.004672	0.000000	0.871604
30	0.002487	0.067355	0.004811	0.000000	0.925348
31	0.002661	0.067349	0.004811	0.000000	0.925180
32	0.002851	0.067343	0.004810	0.000000	0.924997
33	0.003060	0.067336	0.004810	0.000000	0.924796
34	0.003289	0.067328	0.004809	0.000000	0.924575
35	0.003539	0.067319	0.004809	0.000000	0.924334
36	0.003814	0.067310	0.004808	0.000000	0.924069
37	0.004115	0.067300	0.004807	0.000000	0.923779
38	0.004444	0.067288	0.004806	0.000000	0.923462
39	0.004806	0.067276	0.004805	0.000000	0.923114
40	0.005333	0.019700	0.004925	0.000000	0.970042
41	0.005778	0.019695	0.004924	0.000000	0.969603
42	0.006265	0.019691	0.004923	0.000000	0.969122
43	0.006798	0.019685	0.004921	0.000000	0.968595
44	0.007383	0.019680	0.004920	0.000000	0.968018
45	0.008023	0.019673	0.004918	0.000000	0.967386
46	0.008724	0.019666	0.004917	0.000000	0.966694
47	0.009491	0.019659	0.004915	0.000000	0.965935
48	0.010332	0.019651	0.004913	0.000000	0.965105
49	0.011253	0.019642	0.004910	0.000000	0.964196
50	0.012261	0.019632	0.004908	0.000000	0.963200
51	0.013364	0.019621	0.004905	0.000000	0.962110
52	0.014572	0.019609	0.004902	0.000000	0.960917
53	0.015894	0.019596	0.004899	0.000000	0.959611
54	0.017342	0.019582	0.004895	0.000000	0.958182
55$^-$	0.000000	0.000000	0.000000	0.040484	0.670727
55$^+$	0.018549	0.000000	0.004796	0.057548	0.919108
56	0.020247	0.000000	0.004791	0.057498	0.917464
57	0.022105	0.000000	0.004787	0.057443	0.915665
58	0.024138	0.000000	0.004782	0.057383	0.913698
59	0.026360	0.000000	0.004776	0.057318	0.911547
60	0.028790	0.000000	0.004771	0.057246	0.909195
61	0.031446	0.000000	0.004764	0.057168	0.906624
62	0.034348	0.000000	0.004757	0.057082	0.903814
63	0.037519	0.000000	0.004749	0.056988	0.900745
64	0.040982	0.000000	0.004740	0.056885	0.897394
65$^-$	0.000000	0.000000	0.000000	1.000000	0.000000

Table 3.2 Multiple decrement table

Age x	$l_x^{(\tau)}$	$d_x^{(d)}$	$d_x^{(w)}$	$d_x^{(i)}$	$d_x^{(r)}$
25	100000.00	176.74	12149.68	467.30	0.00
26	87206.41	163.44	10594.72	407.49	0.00
27	76040.88	151.41	9237.66	355.29	0.00
28	66296.61	140.51	8053.37	309.74	0.00
29	57793.08	130.60	7019.89	270.00	0.00
30	50372.66	125.27	3392.85	242.35	0.00
31	46612.23	124.02	3139.29	224.24	0.00
32	43124.72	122.96	2904.13	207.44	0.00
33	39890.23	122.06	2686.03	191.86	0.00
34	36890.31	121.32	2483.74	177.41	0.00
35	34107.87	120.72	2296.11	164.01	0.00
36	31527.06	120.24	2122.08	151.58	0.00
37	29133.19	119.87	1960.65	140.05	0.00
38	26912.64	119.61	1810.91	129.35	0.00
39	24852.79	119.43	1672.00	119.43	0.00
40	22941.96	122.36	451.95	112.99	0.00
41	22254.67	128.59	438.37	109.58	0.00
42	21578.19	135.19	424.87	106.22	0.00
43	20911.90	142.17	411.66	102.91	0.00
44	20255.17	149.54	398.62	99.65	0.00
45	19607.37	157.30	385.74	96.44	0.00
46	18967.89	165.47	373.03	93.26	0.00
47	18336.14	174.03	360.47	90.12	0.00
48	17711.53	183.00	348.04	87.01	0.00
49	17093.48	192.35	335.74	83.94	0.00
50	16481.46	202.07	323.56	80.89	0.00
51	15874.95	212.15	311.48	77.87	0.00
52	15273.45	222.57	299.50	74.87	0.00
53	14676.52	233.28	287.60	71.90	0.00
54	14083.75	244.24	275.78	68.95	0.00
55^-	13494.80	0.00	0.00	0.00	4048.44
55^+	9446.36	175.22	0.00	45.30	543.62
56	8682.22	175.79	0.00	41.60	499.21
57	7965.63	176.08	0.00	38.13	457.57
58	7293.85	176.06	0.00	34.88	418.55
59	6664.37	175.67	0.00	31.83	381.99
60	6074.89	174.89	0.00	28.98	347.76
61	5523.25	173.68	0.00	26.31	315.75
62	5007.51	172.00	0.00	23.82	285.84
63	4525.86	169.80	0.00	21.49	257.92
64	4076.65	167.07	0.00	19.33	231.90
65^-	3658.36	0.00	0.00	0.00	3658.36
Total	–	6378.75	76949.48	5225.79	11446.90
h_j	–	0.063792	0.769488	0.052257	0.114469

Example 3.2.2 illustrates the advantage of constructing a multiple decrement table given the forces of decrements to compute similar probabilities.

Example 3.2.2 Suppose that a member of the pension plan, joining the plan at age 25, exits the plan by any one of the four causes of exit: withdrawal, disability, retirement, death in service, and age retirement. The probability law governing these decrements is as specified in Example 3.2.1. Calculate the probability that a member of age 40

(i) retires at exact age 65;
(ii) exits the plan due to (a) death in service, (b) withdrawal, (c) disability, and (d) age-service retirement.

Solution We have prepared multiple decrement table, Table 3.2, corresponding to the specified forces of decrement. Using the expected number of decrements and survivors as displayed in Table 3.2, we compute these probabilities as follows:

(i) The probability that a member of age 40 retires at exact age 65 is

$$_{25}p_{40}^{(\tau)} = l_{65}^{(\tau)} / l_{40}^{(\tau)} = 3658.36/22941.96 = 0.1595.$$

(ii) (40) will exit the plan due to death in service if death occurs in any one of the next 25 years. This probability can be expressed as

$$_{25}q_{40}^{(d)} = q_{40}^{(d)} + p_{40}^{(\tau)}q_{41}^{(d)} + _{2}p_{40}^{(\tau)}q_{42}^{(d)} + \cdots + _{24}p_{40}^{(\tau)}q_{64}^{(d)}$$

$$= \frac{d_{40}^{(d)}}{l_{40}^{(\tau)}} + \frac{l_{41}^{(\tau)}}{l_{40}^{(\tau)}}\frac{d_{41}^{(d)}}{l_{41}^{(\tau)}} + \cdots + \frac{l_{64}^{(\tau)}}{l_{40}^{(\tau)}}\frac{d_{64}^{(d)}}{l_{64}^{(\tau)}}$$

$$= \frac{d_{40}^{(d)}}{l_{40}^{(\tau)}} + \frac{d_{41}^{(d)}}{l_{40}^{(\tau)}} + \cdots \frac{d_{64}^{(d)}}{l_{40}^{(\tau)}}$$

$$= _{25}d_{40}^{(d)} / l_{40}^{(\tau)}.$$

Thus the probability that a member of age 40 exits the plan due to death in service is

$$_{25}q_{40}^{(d)} = _{25}d_{40}^{(d)} / l_{40}^{(\tau)} = 4400.57/22941.96 = 0.1918.$$

Similarly, the probability that a member of age 40 withdraws is

$$_{15}q_{40}^{(w)} = _{15}d_{40}^{(w)} / l_{40}^{(\tau)} = 5426.41/22941.96 = 0.2365.$$

The probability that a member of age 40 exits the plan due to disability is

$$_{25}q_{40}^{(i)} = _{25}d_{40}^{(i)} / l_{40}^{(\tau)} = 1668.27/22941.96 - 0.0727.$$

The probability that a member of age 40 exits the plan due to age-service retirement is

$$_{25}q_{40}^{(r)} = {}_{25}d_{40}^{(r)}/l_{40}^{(\tau)} = 11446.90/22941.96 = 0.4990.$$

It is to be noted that four probabilities computed in (ii) add up to 1, as it should be in view of the fact that (40) will exit the plan in next 25 years either due to one of the four causes. More precisely, $P[J(40) = 1] = 0.1918$, $P[J(40) = 2] = 0.2365$, $P[J(40) = 3] = 0.0727$, and $P[J(40) = 4] = 0.4990$.

A multiple decrement table specifying the expected number of decrements and expected number of active members or the probabilities of decrement and survival probabilities for various ages, starting from some minimum age to some maximum age, is referred to as a service table in the pension funding context. Table 3.3 presents a hypothetical service table. In Table 3.3, the age of entry is 25, the minimum age for service-age retirement is 55, and the age for mandatory retirement is 65. Withdrawal or disability retirement after 55 are treated as age-service retirement. The table displays the expected number of survivors and expected number of decrements due to four causes. Here $l_{65-}^{(\tau)} = 8511$. Thus, 8511 members remain in service up to 65. Since 65 is the mandatory age of retirement, all those who remain in service up to 65 retire at 65. Hence, $d_{65}^{(r)} = 8511$. Superscript $-$ to 65 indicates the beginning of the age interval (65–66).

The last column of Table 3.3 displays the salary scale function w_x, which is defined later, after Table 3.5.

The multiple decrement table, displayed in Table 3.4, specifying the decrement probabilities and the survival probabilities, can be obtained from this service table using the following formulas:

$$q_x^{(d)} = \frac{d_x^{(d)}}{l_x^{(\tau)}}, \qquad q_x^{(w)} = \frac{d_x^{(w)}}{l_x^{(\tau)}}, \qquad q_x^{(i)} = \frac{d_x^{(i)}}{l_x^{(\tau)}}, \qquad q_x^{(r)} = \frac{d_x^{(r)}}{l_x^{(\tau)}}$$

$$\text{and} \quad p_x^{(\tau)} = 1 - \left(q_x^{(d)} + q_x^{(w)} + q_x^{(i)} + q_x^{(r)}\right) \quad \text{or} \quad p_x^{(\tau)} = \frac{l_{x+1}^{(\tau)}}{l_x^{(\tau)}}.$$

Since 65 is the mandatory age of retirement, $q_{65}^{(r)} = \frac{d_{65}^{(r)}}{l_{65-}^{(\tau)}} = 1$.

We note from Tables 3.3 and 3.4 that in the early years of service, withdrawal rates are high and after 10 years withdrawal rates tend to be low. Such a pattern is common in practice. With any employment, there is a mandatory age for retirement. As a consequence, in the service table, $l_x^{(\tau)}$ is 0 after certain age. In Table 3.3, $l_{65}^{(\tau)}$ is 8511. As 65 is a mandatory age of retirement, all 8511 members retire, and hence $l_x^{(\tau)}$ is 0 after age 65. Table 3.4 displays the proportion of exits from active service due to various modes of exit in each unit age interval.

Our main aim is to decide on the pension fund at a specified time. The actuarial present value of benefit due to various causes is the main input for the pension fund. So given the decrement pattern, either in terms of the forces of decrement or in terms

Table 3.3 Service table

Age x	$l_x^{(\tau)}$	$d_x^{(d)}$	$d_x^{(w)}$	$d_x^{(i)}$	$d_x^{(r)}$	w_x
25	100000	22	18600	–	–	1
26	81378	23	13457	–	–	1.05
27	67898	25	9756	–	–	1.11
28	58117	26	5854	–	–	1.17
29	52237	28	3875	–	–	1.23
30	48334	31	2576	36	–	1.30
31	45691	32	1879	38	–	1.36
32	43742	34	1478	41	–	1.44
33	42189	36	1206	43	–	1.51
34	40904	38	967	45	–	1.60
35	39854	39	834	46	–	1.68
36	38935	41	733	47		1.77
37	38114	44	624	49	–	1.87
38	37397	45	547	51	–	1.96
39	36754	47	478	52	–	2.07
40	36177	48	462	66	–	2.18
41	35601	50	412	56	–	2.29
42	35083	54	402	53	–	2.42
43	34574	56	376	45	–	2.55
44	34097	61	334	43	–	2.68
45	33659	66	287	60	–	2.82
46	33246	69	278	87	–	2.97
47	32812	74	265	93	–	3.13
48	32380	85	254	112	–	3.30
49	31929	92	220	121	–	3.46
50	31496	124	214	132	–	3.65
51	31026	136	180	143	–	3.84
52	30567	143	175	156	–	4.03
53	30093	153	145	187	–	4.24
54	29608	178	125	215	–	4.45
55	29090	195	–	–	2566	4.67
56	26329	212	–	–	2096	4.89
57	24021	214	–	–	2143	5.12
58	21664	234	–	–	1678	5.35
59	19752	245	–	–	1782	5.58
60	17725	324	–	–	1985	5.80
61	15416	423	–	–	1560	6.03
62	13433	523	–	–	1245	6.27
63	11665	634	–	–	987	6.52
64	10044	744	–	–	789	6.79
65^-	8511	–	–	–	8511	–

Table 3.4 Multiple decrement table

Age x	$q_x^{(d)}$	$q_x^{(w)}$	$q_x^{(i)}$	$q_x^{(r)}$	$p_x^{(\tau)}$
25	0.00022	0.18600	0.00000	0.00000	0.81378
26	0.00028	0.16536	0.00000	0.00000	0.83435
27	0.00037	0.14369	0.00000	0.00000	0.85595
28	0.00045	0.10073	0.00000	0.00000	0.89882
29	0.00054	0.07418	0.00000	0.00000	0.92528
30	0.00064	0.05330	0.00074	0.00000	0.94532
31	0.00070	0.04112	0.00083	0.00000	0.95734
32	0.00078	0.03379	0.00094	0.00000	0.96450
33	0.00085	0.02859	0.00102	0.00000	0.96954
34	0.00093	0.02364	0.00110	0.00000	0.97433
35	0.00098	0.02093	0.00115	0.00000	0.97694
36	0.00105	0.01883	0.00121	0.00000	0.97891
37	0.00115	0.01637	0.00129	0.00000	0.98119
38	0.00120	0.01463	0.00136	0.00000	0.98281
39	0.00128	0.01301	0.00141	0.00000	0.98430
40	0.00133	0.01277	0.00182	0.00000	0.98408
41	0.00140	0.01157	0.00157	0.00000	0.98545
42	0.00154	0.01146	0.00151	0.00000	0.98549
43	0.00162	0.01088	0.00130	0.00000	0.98620
44	0.00179	0.00980	0.00126	0.00000	0.98715
45	0.00196	0.00853	0.00178	0.00000	0.98773
46	0.00208	0.00836	0.00262	0.00000	0.98695
47	0.00226	0.00808	0.00283	0.00000	0.98683
48	0.00263	0.00784	0.00346	0.00000	0.98607
49	0.00288	0.00689	0.00379	0.00000	0.98644
50	0.00394	0.00679	0.00419	0.00000	0.98508
51	0.00438	0.00580	0.00461	0.00000	0.98521
52	0.00468	0.00573	0.00510	0.00000	0.98449
53	0.00508	0.00482	0.00621	0.00000	0.98388
54	0.00601	0.00422	0.00726	0.00000	0.98250
55	0.00670	0.00000	0.00000	0.08821	0.90509
56	0.00805	0.00000	0.00000	0.07961	0.91234
57	0.00891	0.00000	0.00000	0.08921	0.90188
58	0.01080	0.00000	0.00000	0.07746	0.91174
59	0.01240	0.00000	0.00000	0.09022	0.89738
60	0.01828	0.00000	0.00000	0.11199	0.86973
61	0.02744	0.00000	0.00000	0.10119	0.87137
62	0.03893	0.00000	0.00000	0.09268	0.86838
63	0.05435	0.00000	0.00000	0.08461	0.86104
64	0.07407	0.00000	0.00000	0.07855	0.84737
65^-	0.00000	0.00000	0.00000	1.00000	0.00000

of the survival and decrement probabilities or in terms of the expected number of survivors and expected number of decrements due to various causes, we proceed to find the joint distribution of the random variable $K(x) \equiv K$, the curtate-future time until decrement of (x), and a random variable $J(x) \equiv J$ describing the cause of decrement. The random variable K denotes the complete years of service before exit from the status of active service. As derived in Chap. 1, the joint probability mass function of K and J is given by

$$P[K = k, J = j] = p(k, j) = {}_k p_x^{(\tau)} q_{x+k}^{(j)}.$$

From the multiple decrement table displayed in Table 3.4, we can obtain the joint probability mass function of K and J and the marginal probability mass function of K and J. This is useful in computing the actuarial present value of the benefits, as discussed in the next section. For convenience of future reference, we introduce the following notation for the joint and marginal probability mass function of K and J:

$$p(k, d) = P[K = k, J = 1], \qquad p(k, w) = P[K = k, J = 2],$$
$$p(k, i) = P[K = k, J = 3], \qquad p(k, r) = P[K = k, J = 4], \quad \text{and}$$
$$p_k = P[K = k] = {}_k p_x^{(\tau)} q_{x+k}^{(\tau)}, \quad k = 0, 1, \dots.$$

Further, the marginal distribution of J given by

$$h_j = P[J = j] = \sum_{k \ge 0} P[K = k, J = j], \quad j = 1, 2, 3, 4.$$

We find the joint distribution of $K(25) \equiv K$ and $J(25) \equiv J$, and the marginal distribution of K and J from the multiple decrement model specified in Table 3.4. The following set of R commands computes all these quantities. The results are displayed in Table 3.5. Suppose that Table 3.4 is stored on D drive as mdt.txt file. We begin with importing the file to R console.

```
u <- read.table("D://mdt.txt", header=T);
u1 <- u[, 2]   # q_x^(d);   u2 <- u[, 3]   # q_x^(w);
u3 <- u[, 4]   # q_x^(i);   u4 <- u[, 5]   # q_x^(r);
u5 <- u[, 6]   # p_x^(τ);
p <- c(1, 2:40)   # dummy vector to store _k p25^(τ)
                  # for k = 0 to 40;
for (i in 2:40)
  {
  p[i] <- p[i-1]*u5[i-1]
  }
p   # a vector of _k p25^(τ) for k = 0 to 40;
qd <- p*u1   # _k p25^(τ) q25+k^(d);
qw <- p*u2   # _k p25^(τ) q25+k^(w);
```

```
qi <- p*u3       # k p_25^(τ) q_25+k^(i);
qr <- p*u4       # k p_25^(τ) q_25+k^(r);
d <- data.frame(qd, qw, qi, qr);
h <- rowSums(d)    # p_k for k = 0, 1, ..., 40;
h1 <- sum(h);
h1  # it is approximately 1;
m <- colSums(d  # h_j for j = 1, 2, 3, 4;
m1 <- sum(m);
m1 # it is approximately 1 and same as h1;
k <- 0:40;
d2 <- data.frame(k, qd, qw, qi, qr, h);   d2  # Table 3.5;
```

Table 3.5 displays the joint and marginal probability distribution of K and J, with the last row and the last column displaying the marginal distribution of J and K, respectively.

In Table 3.5, $p(39, r) = P[K(25) = 39, J(25) = 4]$ indicates the chance of retirement in the age interval (64–65), while

$$p(40, r) = P\left[K(25) = 40, J(25) = 4\right] = {}_{40}p_{25}^{(\tau)} q_{65}^{(r)} = 0.08511 \times 1 = 0.08511$$

indicates the chance of retirement at exact age 65.

It is to be noted that the probability distributions displayed in Table 3.5 are in context of the service Table 3.3. Thus for the given service table, the probability that death occurs in active service is 0.05648, the probability of withdrawal is 0.66994, quite high, the probability of disability is 0.02016, and the probability of retirement due to age service is 0.25342. Further, from the last column we observe that the probability of complete years of service is high for $k = 0$ and $k = 1$, mainly due to high chance of withdrawals in early years.

Example 3.2.3 On the basis of service table given in Table 3.3, find the probability that (40)

(i) withdraws,
(ii) becomes disabled,
(iii) dies in active service, and
(iv) retires.

Solution (i) The solution is similar to that of Example 3.2.2. (40) may withdraw up to age 55, so the probability that (40) withdraws is given by

$$\sum_{40}^{54} d_x^{(w)} / l_{40}^{(\tau)} = 4129/36177 = 0.114133.$$

Table 3.5 Joint and marginal distributions of K and J

k	$p(k,d)$	$p(k,w)$	$p(k,i)$	$p(k,r)$	p_k
0	0.00022	0.18600	0.00000	0.00000	0.18622
1	0.00023	0.13457	0.00000	0.00000	0.13479
2	0.00025	0.09756	0.00000	0.00000	0.09781
3	0.00026	0.05854	0.00000	0.00000	0.05880
4	0.00028	0.03875	0.00000	0.00000	0.03903
5	0.00031	0.02576	0.00036	0.00000	0.02643
6	0.00032	0.01879	0.00038	0.00000	0.01949
7	0.00034	0.01478	0.00041	0.00000	0.01553
8	0.00036	0.01206	0.00043	0.00000	0.01285
9	0.00038	0.00967	0.00045	0.00000	0.01050
10	0.00039	0.00834	0.00046	0.00000	0.00919
11	0.00041	0.00733	0.00047	0.00000	0.00821
12	0.00044	0.00624	0.00049	0.00000	0.00717
13	0.00045	0.00547	0.00051	0.00000	0.00643
14	0.00047	0.00478	0.00052	0.00000	0.00577
15	0.00048	0.00462	0.00066	0.00000	0.00576
16	0.00050	0.00412	0.00056	0.00000	0.00518
17	0.00054	0.00402	0.00053	0.00000	0.00509
18	0.00056	0.00376	0.00045	0.00000	0.00477
19	0.00061	0.00334	0.00043	0.00000	0.00438
20	0.00066	0.00287	0.00060	0.00000	0.00413
21	0.00069	0.00278	0.00087	0.00000	0.00434
22	0.00074	0.00265	0.00093	0.00000	0.00432
23	0.00085	0.00254	0.00112	0.00000	0.00451
24	0.00092	0.00220	0.00121	0.00000	0.00433
25	0.00124	0.00214	0.00132	0.00000	0.00470
26	0.00136	0.00180	0.00143	0.00000	0.00459
27	0.00143	0.00175	0.00156	0.00000	0.00474
28	0.00153	0.00145	0.00187	0.00000	0.00485
29	0.00178	0.00125	0.00215	0.00000	0.00518
30	0.00195	0.00000	0.00000	0.02566	0.02761
31	0.00212	0.00000	0.00000	0.02096	0.02308
32	0.00214	0.00000	0.00000	0.02143	0.02357
33	0.00234	0.00000	0.00000	0.01678	0.01912
34	0.00245	0.00000	0.00000	0.01782	0.02027
35	0.00324	0.00000	0.00000	0.01985	0.02309
36	0.00423	0.00000	0.00000	0.01560	0.01983
37	0.00523	0.00000	0.00000	0.01245	0.01768
38	0.00634	0.00000	0.00000	0.00987	0.01621
39	0.00744	0.00000	0.00000	0.00789	0.01533
40	0.00000	0.00000	0.00000	0.08511	0.08511

h_j	h_d	h_w	h_i	h_r	Total
--	0.05648	0.66994	0.02016	0.25342	1

Similarly, the probability that (40) becomes disabled is given by

$$\sum_{40}^{54} d_x^{(i)} / l_{40}^{(\tau)} = 1569/36177 = 0.04337.$$

The probability that (40) dies in service is given by

$$\sum_{40}^{64} d_x^{(d)} / l_{40}^{(\tau)} = 5137/36177 = 0.141996.$$

The probability that (40) retires is

$$\sum_{55}^{64} d_x^{(r)} / l_{40}^{(\tau)} = 25342/36177 = 0.7005.$$

It is to be noted that these four probabilities add to 1 as these are the probabilities of exit due to four causes from the plan for (40), in the context of service Table 3.3.

So far we have discussed one aspect, namely the decrement model, for the group of active members, to find the actuarial present values of the benefits. To find these values, we have to define clearly the benefits for various types of decrement. We begin with the benefit function for the retirement for age-service. In most of the pension plans, the rate of retirement income is defined by a formula. A pension plan that defines a benefit for an employee upon retirement is known as a defined benefit pension plan. Traditionally, retirement plans have been administered by institutions which exist specifically for that purpose, by large businesses, or, for government workers, by the government itself. A typical form of defined benefit plan is the final salary plan, under which the pension paid is equal to the number of years worked, multiplied by the member's salary at or near retirement, multiplied by a suitable fraction. Such an amount is available as an initial monthly pension. In this plan the benefit income rate involves a function of the average salary at or near retirement. Sponsor contributions are also expressed as a percentage of salary. So in both the cases it is essential to estimate the future salaries. The following are some salary functions needed for the estimation. Suppose that the participant of the pension plan joins the service at complete age a and the employer wishes to find the actuarial present value of the retirement benefit after h years that is, at age $(a + h)$, when the employee is in active service. We assume that both a and h are integers.

1. $(AS)_{a+h}$ is the actual annual salary rate at age $a + h$ for a participant who entered at age a and has attained age $a + h$.
2. $(ES)_{a+h+t}$ is the estimated annual salary rate at age $a + h + t$.
3. w_x is the salary scale function, which reflects merit and seniority increases in salary and increases due to inflation. For example, suppose that the deterministic model for salary scale function is specified as $w_x = (1.05)^{x-25} u_x$, where u_x represents the progression of salary due to individual merit and experience, and 5 % accumulation factor to allow for the long-term inflation and of increases

in productivity of all members of the plan. The initial value of w_x at $x = 25$ is chosen arbitrarily as 1, just as $l_{25}^{(\tau)}$ is taken as 100000. The w_x function is usually assumed to be a step function, with constant level throughout any given year of age. Such modeling of the salary scale function is similar to the exponential growth, $(1+i) = e^\delta$, used for continuous compounding interest. The salary scale function decides the size of the salary progression, and it has dramatic impact on the level of benefit and hence on the projected cost of the plan. The last column of Table 3.3 presents a typical salary scale function. Thus, $w_{40} = 2.18$ indicates that for the employee of age 40, the salary scale for the next one year is 2.18, and $w_{64} = 6.79$ indicates that for the employee of age 64, the salary scale for the last year of employment is 6.79. With the salary scale function, an estimated annual salary rate can be obtained from the actual annual salary rate by the rule of threes as

$$(ES)_{a+h+t} = (AS)_{a+h} \frac{w_{a+h+t}}{w_{a+h}}.$$

As an illustration, in the following example we find the estimated salary of (40) for the next 25 years, using the salary scale function as given in Table 3.3.

Example 3.2.4 Suppose that an individual is hired at age of 25 and is in active service at 40 with annual salary Rs 500000/-. Using the salary scale function as given in Table 3.3, find the estimated salary for the next 25 years.

Solution To find the estimated salaries, we use the formula

$$(ES)_{40+t} = (AS)_{40} \frac{w_{40+t}}{w_{40}} = 500000 \frac{w_{40+t}}{w_{40}}.$$

The estimated salaries are reported in Table 3.6.

From Table 3.6 we note that if the annual salary at age 40 is Rs 500000/-, then according to the salary scale function w_x as given in Table 3.3, the estimated salary during (50–51) is Rs 837156.00/- and during (64–65) and hence at the age of retirement, it is Rs 1557339.40/-.

The projected salary is useful to estimate the initial benefit level for a pension plan. Toward it we define a function $R(a, h, t)$ as the projected annual benefit rate to commence at age $a + h + t$ for a participant who entered at age a, is in service at age $a + h$, and retires at $a + h + t$. Suppose that the pension benefit is the monthly life annuity due. If the pension benefit rate remains level during payout, then the actuarial present value of the benefit at time of retirement is given by $R(a, h, t)\ddot{a}_{a+h+t}^{(12)r}$. The superscript r in $\ddot{a}_{a+h+t}^{(12)r}$ indicates that the actuarial present value of the life annuity due is obtained using the life table for retired lives. In practice the pension payment does not remain the same for all the future years but increases periodically, usually annually. To incorporate this increase, let $h(x)$ denote an adjustment factor applied to the initial pension payment rate $R(a, h, t)$ for those who retired $x-y$ years ago, y being the age of retirement. $h(x)$ is usually determined using the consumer

Table 3.6 Estimated salary

t	Age	$(ES)_{40+t}$	t	Age	$(ES)_{40+t}$
1	41–42	525229.40	13	53–54	972477.10
2	42–43	555045.90	14	54–55	1020642.20
3	43–44	584862.40	15	55–56	1071100.90
4	44–45	614678.90	16	56–57	1121559.60
5	45–46	646789.00	17	57–58	1174311.90
6	46–47	681192.70	18	58–59	1227064.20
7	47–48	717889.90	19	59–60	1279816.50
8	48–49	756880.70	20	60–61	1330275.20
9	49–50	793578.00	21	61–62	1383027.50
10	50–51	837156.00	22	62–63	1438073.40
11	51–52	880733.90	23	63–64	1495412.80
12	52–53	924311.90	24	64–65	1557339.40

price index or retail price index. As an example, suppose that $h(x) = \exp[\eta(x - y)]$, where η is a constant rate of increase, possibly related to the expected inflation rate. With such an adjustment function $h(x)$, the annual increase in the pension payment is at the rate of e^{η}. We obtain the expression for such an increasing annuity, with $h(x) = \exp[\eta(x - y)]$. Let δ denote the force of interest, and $\bar{a}_y^{\eta}$ denote the actuarial present value of the life annuity at age y of retirement, of the pension benefit which takes into account the increase in pension as governed by the function $h(x)$. The superscript η in $\bar{a}_y^{\eta}$ indicates the role of the rate η. By definition,

$$\bar{a}_y^{\eta} = \int_0^{\infty} v^t h(t) \,_t p_y \, dt = \int_0^{\infty} e^{-\delta t} e^{\eta t} \,_t p_y \, dt$$

$$= \int_0^{\infty} e^{-(\delta - \eta)t} \,_t p_y \, dt = \int_0^{\infty} e^{-\delta' t} \,_t p_y \, dt$$

$$= \bar{a}_y(\delta'),$$

where $\delta' = \delta - \eta$, that is, $\bar{a}_y^{\eta}$ is the annuity function $\bar{a}_y$ with force of interest $\delta' = \delta - \eta$. Analogously it can be proved that

$$\ddot{a}_y^{\eta} = \sum_{k=0}^{\infty} v^k h(k) \,_k p_y = \sum_{k=0}^{\infty} e^{-\delta k} e^{\eta k} \,_k p_y = \sum_{k=0}^{\infty} v'^k \,_k p_y = \ddot{a}_y^{\eta}(v'),$$

where $v' = e^{-\delta'}$. Once we get $\ddot{a}_y^{\eta}$, we can find $\ddot{a}_y^{\eta(12)}$ by using the relation

$$\ddot{a}_x^{(12)} = \alpha(12)\ddot{a}_x - \beta(12), \quad \text{where } \alpha(12) = \frac{id}{i^{(12)}d^{(12)}} \text{ and } \beta(12) = \frac{i - i^{(12)}}{i^{(12)}d^{(12)}}.$$

The nominal interest rate $i^{(12)}$ and nominal discount rate $d^{(12)}$ are given by

$$i^{(12)} = 12\{(1+i)^{1/12} - 1\} \quad \text{and} \quad d^{(12)} = 12(1 - v^{1/12}).$$

Thus, $\ddot{a}_y^{(12)\eta}$ is $\ddot{a}_y^{(12)}$ with force of interest δ'. In further discussion we use $\ddot{a}_y^{(12)\eta}$ to find the actuarial present values of benefits.

We now proceed to define the most frequently used formulas for $R(a, h, t)$ for integer values of a, h, and t. The first two do not take into account the amount of service, but in the last two, the benefit is proportional to the number of years of service at retirement.

1. Suppose that the individual retires at age $a + h + t$ and the benefit rate for the pension is the fraction d of the final salary rate. d is sometimes referred to as the accrual rate. Then annual initial benefit rate for the pension is projected as

$$R_1(a, h, t) = d(ES)_{a+h+t-1} = d(AS)_{a+h} \frac{w_{a+h+t-1}}{w_{a+h}}.$$

The benefit rate is estimated from the current salary rate at $a + h$.

2. In the first approach, the benefit rate $R_1(a, h, t)$ is a function of only the final salary; instead, more frequently, the final m-year average salary rate is taken into account. m is usually taken as 3 or 5. Thus the benefit rate is defined as the fraction d of the final m-year average salary rate. The 5-year average salary rate, denoted $_5Z_{a+h+t}$, is defined as

$$_5Z_{a+h+t} = \frac{w_{a+h+t-5} + w_{a+h+t-4} + w_{a+h+t-3} + w_{a+h+t-2} + w_{a+h+t-1}}{5}.$$

For example, if a person retires at the age of 60, then the 5-year average salary rate would be the average of salary at ages 55, 56, 57, 58, and 59. The 5-year average salary rate, denoted $_5Z_{60}$, is then defined as

$$_5Z_{60} = \frac{w_{55} + w_{56} + w_{57} + w_{58} + w_{59}}{5}.$$

If $t > 5$, the initial benefit rate depending on the 5-year average salary rate, denoted $R_2(a, h, t)$, is defined as

$$R_2(a, h, t) = \frac{d(AS)_{a+h} \, _5Z_{a+h+t}}{w_{a+h}}.$$

If $t < 5$, the actual salary for some years is taken into account instead of estimated salaries.

3. Often, both the amount of service and the final 5-year salary are taken into account to define the initial benefit rate for pension. In this case, the initial benefit rate, denoted $R_3(a, h, t)$, is given by

$$R_3(a, h, t) = d(h + t) \frac{(AS)_{a+h} \, _5Z_{a+h+t}}{w_{a+h}},$$

where $(h + t)$ denotes the total period of service. Here also, if $t < 5$, the actual salary is taken into account instead of estimated salaries.

4. Sometimes the entire career earnings are taken into account, instead of m years. Thus we have

$$R_4(a, h, t) = d(h + t) \times \text{Average Salary over the Entire Career.}$$

When the entire career earnings are taken into account, salary information is known for the past service and is estimated for the future service.

5. For some employees, the number of years of service is large. Then the formula for $R(a, h, t)$ is appropriately modified for excess years of service. For example,

$$R(a, h, t) = 0.02(h + t)\frac{(AS)_{a+h} \, _5Z_{a+h+t}}{w_{a+h}} \quad \text{if } h + t \le 30 \quad \text{and}$$

$$R(a, h, t) = \left[0.6 + 0.01(h + t)\right]\frac{(AS)_{a+h} \, _5Z_{a+h+t}}{w_{a+h}} \quad \text{if } h + t \ge 30.$$

In this case, the benefit rate is the product of 5-year final average salary and 0.02 times the number of years of service. If the number of years of service exceeds 30 years, then for each exceeding year, there is an additional benefit as 0.01 times number of years of service above 30 years.

In Maharashtra State government initial monthly pension benefit is 0.5 multiplied by the last month's basic pay or average of last 10 months basic pay, whichever is maximum.

The following example illustrates the computation of estimated initial benefit rate starting from age 55, assuming that it is the minimum eligible age for age-service retirement.

Example 3.2.5 Suppose that an individual enters the job at age 25. He is eligible for the retirement benefit if he retires at any age between 55 to 65. Assume that the salary at age 40 is Rs 500000/- and $d = 0.2$ if the benefit rate is fraction d of the final salary rate or fraction d of the 5-year average salary rate. Suppose that $d = 0.007$ if the benefit rate is fraction d of the 5-year average salary rate multiplied by the total years of service. Assume that the salary scale function is as given in the last column of Table 3.3. Find at age 40 the projected annual initial pension benefit for ages 55 to 65.

Solution We use the formulas as derived above to write an R code for the computation of projected annual benefit. We have $R_1(25, 15, t) = 0.2(ES)_{40+t-1}$. Thus, the initial pension benefit if the individual retires at age 55 is the fraction of salary when age 55 is completed. The salary during age (54–55) is estimated as $(ES)_{54}$. Similarly, the estimated salary at the retirement at age 65 is $(ES)_{64}$. We have $R_2(25, 15, t) = \frac{0.02(AS)_{40} \, _5Z_{40+t}}{w_{40}}$. In this approach, the initial pension benefit is the fraction of the average of last 5-years' salary. Thus, to find the initial pension benefit at age 55, we take the average of salary at complete ages 50, 51, 52, 53, and

Table 3.7 Projected annual benefit at retirement

t	Age	$R_1(25, 15, t)$	$R_2(25, 15, t)$	$R_3(25, 15, t)$
15	55	204128.40	185412.80	194683.50
16	56	214220.20	194770.60	211326.10
17	57	224311.90	204403.70	228932.10
18	58	234862.40	214403.70	247636.20
19	59	245412.80	224587.20	267258.70
20	60	255963.30	234954.10	287818.80
21	61	266055.00	245321.10	309104.60
22	62	276605.50	255779.80	331234.90
23	63	287614.70	266330.30	354219.30
24	64	299082.60	277064.20	378192.70
25	65	311467.90	288165.10	403431.20

54. Thus, $_5Z_{40+15} = \frac{w_{50}+w_{51}+w_{52}+w_{53}+w_{54}}{5}$. In the third approach, $R_3(25, 15, t) = \frac{0.007(15+t)(AS)_{40}\,_5Z_{40+t}}{w_{40}}$, where $_5Z_{40+t}$ is computed as in $R_2(25, 15, t)$. Suppose that two columns, the age and salary scale function, of the service table as given in Table 3.3, are stored on D drive as tab delimited Excel file, service.txt. We begin with importing the file to the R console:

```
z <- read.table("D://service.txt", header=T);
u <- z[30:40, 2]
                #salary scale function for ages 54 to 64;
v <- u/z[16, 2]   # w_40+t−1/w_40 for t = 15 to 25;
v1 <- 0.2*500000*v   # R_1(25,15,t);
x <- z[26:40, 2]
                #salary scale function for ages 50 to 64;
y <- 1:11   #a dummy vector to store the 5-year averages
            #for ages 55 to 65;
for (i in 1:11)
  {
  y[i] <- (x[i]+x[i+1]+x[i+2]+x[i+3]+x[i+4])/5
  }
y  #a vector of the 5-year averages for ages 55 to 64;
y1 <- 0.2*500000*y/z[16, 2]   # R_2(25,15,t);
t <- 15:25;
y2 <- (15+t)*0.007*500000*y/z[16, 2]   # R_3(25,15,t);
d <- round(data.frame(v1, y1, y2), 2);
a <- 55:65;
d1 <- data.frame(t, a, d);
d1   #Table 3.7;
```

Table 3.7 displays the computations of the annual initial pension benefit for three different types of formulas, for the given salary scale function and the specified values of fraction d. Thus for the given setup, we note that the annual initial pension benefit if individual retires at age 55 is 204128.40, according to formula for $R_1(25, 15, 15)$ if the estimated annual salary of the individual at age 55 is 1071100.90 (as obtained in Table 3.6). The annual initial pension benefit amount according to formula for $R_2(a, h, t)$ (fourth column) is less than that corresponding to $R_1(25, 15, t)$ as $R_2(25, 15, t)$ is a function of the 5-year average salary. The annual initial pension benefit amounts given by $R_3(25, 15, t)$ increase fast when the duration of service is larger than 36 years, that is, after age 61.

The multiple decrement model, the estimated salary, and the projected initial pension benefit are the important components of the actuarial present value of the benefits when the decrement from the group is due to one of the four causes. As discussed in Chaps. 1 and 2, the actuarial present value A of the benefits is given by the following formula:

$$A = E\big(b^{(J)}(T)v^T\big) = E\big[E\big(b^{(J)}(T)v^T\big)|J\big] = \sum_{j=1}^{4} E\big(b^{(j)}(T)v^T | J = j\big)P[J = j]$$

$$= \sum_{j=1}^{4}\Big[\int_0^\infty b^{(j)}(t)v^t f(t, j)/h_j \, dt\Big]h_j = \sum_{j=1}^{4}\int_0^\infty b^{(j)}(t)v^t f(t, j) \, dt.$$

The information about the actuarial present value of all the benefits will be helpful to the employer to pay the appropriate contribution in the pension fund. We begin with age-service retirement benefit. We assume that the benefit is the life annuity due at the initial rate of $R(a, h, t)$, increasing by a factor e^η per annum. $R(a, h, t)$ may be any one of the forms defined above. The annuity function $\ddot{a}^{(12)\eta}_{a+h+t}$ denotes the actuarial present value of the monthly life annuity due with unit benefit to begin at age $a + h + t$, incorporating the annual increase e^η. We assume that a life table appropriate for retired lives is used to find the annuity function. Let α denote the minimum eligible age for retirement, and β denote the mandatory age of retirement. Suppose that the individual enters the service at age a, is in active service at age $a + h$, and retires at age $a + h + t$. Then $\alpha \le a + h + t \le \beta$. We obtain the actuarial present value of the retirement benefit annuity both in continuous and discrete setups. For the continuous setup, to find the expected values, we need the force of mortality function. Let $\mu_x^{(w)}(t)$, $\mu_x^{(d)}(t)$, $\mu_x^{(i)}(t)$, and $\mu_x^{(r)}(t)$ denote the forces of decrement at age $(x + t)$ due to withdrawal from service, death in service, retirement for disability, and age-service retirement, respectively. These are assumed to be continuous functions at most ages. The discontinuity occurs at the first eligible age for retirement. Sometimes in the calculation of actuarial present values, a select survival model is used to define the probability of decrement due to withdrawal or disability, with some select period. Survival and decrement probabilities will be appropriately modified in that case. In the following we assume an aggregate survival model. The actuarial present value of the age-service retirement benefit annuity at $a+h+t$ is $b(t) = R(a, h, t)\ddot{a}^{(12)\eta}_{a+h+t}$ if the individual of age $a+h$ retires at $a+h+t$, that is, if $\alpha - a - h \le t \le \beta - a - h$. It is 0 for all other t values. We want to find the actuarial present value of $b(t)$ at $a + h$, so we have to multiply $b(t)$ by v^t. Thus, the actuarial present value at $a + h$, of the retirement benefit annuity, is given by

$$E_r = E\big(b(T)v^T | J = 4\big) = \int_{\alpha-a-h}^{\beta-a-h} v^t \, {}_t p_{a+h}^{(\tau)} \mu_{a+h+t}^{(r)} \frac{1}{h_r} R(a, h, t) \ddot{a}_{a+h+t}^{(12)\eta} \, dt.$$

Here $\frac{1}{h_r} \, {}_t p_{a+h}^{(\tau)} \mu_{a+h+t}^{(r)} \, dt = P[t \leq T(a + h) \leq t + dt | J = 4]$, as defined in Chap. 1. We denote $h_r E_r$ by APV(r). Thus, we have

$$\text{APV}(r) = \int_{\alpha-a-h}^{\beta-a-h} v^t \, {}_t p_{a+h}^{(\tau)} \mu_{a+h+t}^{(r)} R(a, h, t) \ddot{a}_{a+h+t}^{(12)\eta} \, dt.$$

In the continuous-time approach to find the expression for APV(r), we need to know the force of mortality function $\mu_x^{(r)}(t)$. Even if it is known, usually it is not possible to find the explicit value of the integral. Hence, we make some assumptions to compute APV(r). As a first step in the practical computation, we take $T = K + U$, where K is the curtate future life random variable, and the random variable U indicates the fractional age. Thus, we discretize the formula for APV(r) by substituting $t = k + s$, so we get

$$\text{APV}(r) = \sum_{k=\alpha-a-h}^{\beta-a-h} v^k \, {}_k p_{a+h}^{(\tau)} \int_0^1 v^s \, {}_s p_{a+h+k}^{(\tau)} \mu_{a+h+k+s}^{(r)} R(a, h, k + s) \ddot{a}_{a+h+k+s}^{(12)\eta} \, ds.$$

If we further assume that in each year of eligible age, retirements are uniformly distributed, then we have ${}_s p_{a+h+k}^{(\tau)} \mu_{a+h+k+s}^{(r)} = q_{a+h+k}^{(r)}$, and hence APV$(r)$ simplifies to

$$\text{APV}(r) = \sum_{k=\alpha-a-h}^{\beta-a-h} v^k \, {}_k p_{a+h}^{(\tau)} q_{a+h+k}^{(r)} \int_0^1 v^s R(a, h, k + s) \ddot{a}_{a+h+k+s}^{(12)\eta} \, ds.$$

Further simplification can be obtained by the midpoint formula for the integral, which gives

$$\text{APV}(r) = \sum_{k=\alpha-a-h}^{\beta-a-h} v^k \, {}_k p_{a+h}^{(\tau)} q_{a+h+k}^{(r)} v^{0.5} R(a, h, k + 0.5) \ddot{a}_{a+h+k+0.5}^{(12)\eta}.$$

When the continuous setup is discretized, the annuity payments can be viewed as the payments made in the middle of each year. It is to be noted that the expression in continuous setup is essentially converted into a discrete version in terms of summation under the assumption of uniformity for fractional ages and then using midpoint formula. Another approach is to find an expression for APV(r) in a discrete setup, that is, using the joint probability distribution of K and J. In the following, we obtain APV(r) adopting this approach. Suppose that the individual enters the service at age a, is in active service at age $a + h$, and retires at age $a + h + k$. Here we assume that a, h, and k are all integers. Then $\alpha \le a + h + k \le \beta$. Thus, k varies from $\alpha - a - h$ to $\beta - a - h$. If the member retires at age $a + h + k$, then the initial annual retirement benefit will be $R(a, h, k)$, and it depends on the salary scale up to $a + h + k - 1$. The actuarial present value of the retirement benefit annuity at age $a + h$, denoted DE_r, is then given by

$$DE_r = E\left(b(K)v^K | J = 4\right) \sum_{k=\alpha-a-h}^{\beta-a-h} v^k \, {}_k p_{a+h}^{(\tau)} q_{a+h+k}^{(r)} \frac{1}{h_r} R(a, h, k) \ddot{a}_{a+h+k}^{(12)\eta}.$$

As in the continuous setup, we denote $h_r DE_r$ by APV(r), and then APV(r) is given by

$$\text{APV}(r) = \sum_{k=\alpha-a-h}^{\beta-a-h} v^k \, {}_k p_{a+h}^{(\tau)} q_{a+h+k}^{(r)} R(a, h, k) \ddot{a}_{a+h+k}^{(12)\eta}.$$

Here we implicitly assume that the retirements take place at the end of year $a + h + k$, with probability ${}_k p_{a+h}^{(\tau)} q_{a+h+k}^{(r)}$ for $k = \alpha - a - h$ to $\beta - a - h$. Thus the retirement benefit starts at the end of year $a + h + k$, so to find its present value at age $a + h$, the discount factor is v^k. When the multiple decrement table, as displayed in Table 3.4 or Table 3.5, the salary scale function, as given by the last column in Table 3.3, and the annuity values are available, APV(r) can be computed. We illustrate the computation in Example 3.2.6.

Using the steps similar to APV(r), we find the actuarial present values of the benefits corresponding to other modes of decrement. In the case of withdrawal, the employee may be eligible for deferred pension if he has served for some minimum number of years. For example, suppose that an employee with at least 20 years of service is eligible for the retirement benefit if he withdraws. Withdrawal after the minimum eligible age α is treated as retirement. Thus in case of withdrawal, the initial retirement benefit payment is made at age α. Suppose that the individual joins the employer at age a and withdraws from the service between ages $a + 20$ and α; he is then eligible for retirement benefit. Suppose that the current age of the employee is $a + h$ and he withdraws at $a + h + t$; then we have $a + 20 \le a + h + t \le \alpha$, that is, $20 - h \le t \le \alpha - a - h$. The benefit function is $R(a, h, t)$, obtained using any one of the formulas listed above. The benefit will be payable after age α, so the first payment is deferred for $\alpha - a - h - t$ period. The actuarial present value of such withdrawal benefit annuity is obtained using the conditional distribution of T given $J = 2$. We denote it by E_w and then denote $h_w E_w$ by APV(w), which in a continuous case is given by

$$\text{APV}(w) = \int_{20-h}^{\alpha-a-h} v^t \, {}_t p_{a+h}^{(\tau)} \mu_{a+h+t}^{(w)} R(a, h, t) \, {}_{(\alpha-a-h-t)|}\ddot{a}_{a+h+t}^{(12)\eta} \, dt.$$

We can simplify it using the same approach adopted for APV(r). To obtain the expression of APV(w) in discrete setup, suppose that $(a + h)$ withdraws at the end of year $a + h + k$ with probability ${}_k p_{a+h}^{(\tau)} q_{a+h+k}^{(w)}$, the initial retirement benefit depends on the salary at the end of year $a + h + k$, and benefit is not available at the end of year $a + h + k$ but is deferred for $(\alpha - a - h - k)$ years. To obtain its actuarial present value at $a + h$, it is multiplied by v^k. Thus, in discrete setup, $\text{APV}(w)$ is given by

$$\text{APV}(w) = \sum_{k=20-h}^{\alpha-a-h-1} v^k \, {}_k p_{a+h}^{(\tau)} q_{a+h+k}^{(w)} R(a,h,k) \, {}_{(\alpha-a-h-k)|}\ddot{a}_{a+h+k}^{(12)\eta}.$$

Using the identities $_{n|}\ddot{a}_x^{(m)} = {}_n E_x \ddot{a}_{x+n}^{(m)} = v^n \, {}_n p_x \ddot{a}_{x+n}^{(m)}$ and $_k p_x \, _y p_{x+k} = {}_{k+y} p_x$, the expression for $\text{APV}(w)$ gets simplified as follows:

$$\text{APV}(w) = \sum_{k=20-h}^{\alpha-a-h-1} v^k \, {}_k p_{a+h}^{(\tau)} q_{a+h+k}^{(w)} R(a,h,k) \, {}_{(\alpha-a-h-k)|}\ddot{a}_{a+h+k}^{(12)\eta}$$

$$= \sum_{k=20-h}^{\alpha-a-h-1} v^k \, {}_k p_{a+h}^{(\tau)} q_{a+h+k}^{(w)} R(a,h,k) \, {}_{(\alpha-a-h-k)} E_{a+h+k} \ddot{a}_{a+h+k+\alpha-a-h-k}^{(12)\eta}$$

$$= \sum_{k=20-h}^{\alpha-a-h-1} v^k \, {}_k p_{a+h}^{(\tau)} q_{a+h+k}^{(w)} R(a,h,k) v^{\alpha-a-h-k} \, {}_{\alpha-a-h-k} p_{a+h+k} \ddot{a}_{\alpha}^{(12)\eta}$$

$$= v^{\alpha-a-h} \, {}_{\alpha-a-h} p_{a+h} \ddot{a}_{\alpha}^{(12)\eta} \sum_{k=20-h}^{\alpha-a-h-1} q_{a+h+k}^{(w)} R(a,h,k).$$

The last expression of $\text{APV}(w)$ derived above shows that if withdrawal takes place at the end of year $a + h + k$, the payment is deferred till age α, provided that $(a + h)$ survives till α. The benefit will depend on the salary scale at withdrawal.

We now proceed to find the actuarial present value of the disability benefit if disability occurs in active service and hence the individual retires. Suppose that retirement due to disability after age α is treated as retirement due to age-service. Then proceeding on exactly similar lines as for $\text{APV}(w)$ we get the actuarial present value $\text{APV}(i)$ of the disability benefit in continuous setup as follows:

$$\text{APV}(i) = \int_0^{\alpha-a-h} v^t \, {}_t p_{a+h}^{(\tau)} \mu_{a+h+t}^{(i)} R(a,h,t)_{\alpha-a-h-t|} \ddot{a}_{a+h+t}^{(12)\eta} \, dt.$$

In discrete set up $\text{APV}(i)$ is given by

$$\text{APV}(i) = \sum_{k=0}^{\alpha-a-h-1} v^k \, {}_k p_{a+h}^{(\tau)} q_{a+h+k}^{(i)} R(a,h,k) \, {}_{(\alpha-a-h-k)|}\ddot{a}_{a+h+k}^{(12)\eta}.$$

The expression in the discrete setup again gets simplified as follows, using the identities for the deferred annuity as in case of APV(w):

$$\text{APV}(i) = v^{\alpha-a-h}\,_{\alpha-a-h}p_{a+h}\ddot{a}_{\alpha}^{(12)\eta} \sum_{k=0}^{\alpha-a-h-1} q_{a+h+k}^{(i)} R(a,h,k).$$

In practice, the individual may be eligible for disability retirement benefit if he has served for some minimum years of service, in which case lower limit 0 in the above integral and sum will be replaced by the appropriate number. The appropriate modifications in the above formula will give APV(i). In some plans individuals suffering from disability may be eligible for some disability benefits for certain period after disability, depending on the nature of disability. It is to be noted that the actuarial present value of withdrawal benefit or the disability benefit takes the form of the actuarial present value corresponding to benefit in the term life insurance. The last possible mode of exit from the active service is death during active service. In the following, we find the actuarial present value of the death benefit if death occurs in active service. Suppose that the death benefit is thrice the salary rate at the time of death. Suppose that the individual joins the employer at age a, the current age of the employee is $a + h$, and he dies at $a + h + t$, we have $a + h \le a + h + t \le \beta$, that is, $0 \le t \le \beta - a - h$, β being the mandatory age of retirement. The actuarial present value of such death benefit, denoted APV(d), is then given by

$$\text{APV}(d) = \int_0^{\beta-a-h} v^t \,_t p_{a+h}^{(\tau)} \mu_{a+h+t}^{(d)} 3(ES)_{a+h+t}\, dt.$$

The actuarial present value of the death benefit is similar to that in the whole life insurance where the death benefit is a continuous increasing function. We can simplify the expression using the uniformity assumption of death in each unit age interval and finally using the midpoint approximation to the integral. In the discrete setup, suppose that death occurs in the interval $(a + h + k, a + h + k + 1)$; then the benefit is $3(ES)_{a+h+k}$ and is paid at the end of year of death. Hence, in the discrete setup, APV(d) is given by

$$\text{APV}(d) = \sum_{k=0}^{\beta-a-h-1} v^{k+1} \,_k p_{a+h}^{(\tau)} q_{a+h+k}^{(d)} 3(ES)_{a+h+k}.$$

Combining the actuarial present values of the benefits corresponding to all the modes of decrements, we get the actuarial present value, APV, of the total benefit. Let $E(X_j | J = j)$ denote the actuarial present value corresponding to the jth mode of decrement. Then APV is given by

$$\text{APV} = \sum_{j=1}^{4} E(X_j | J = j) P[J = j] = \text{APV}(d) + \text{APV}(w) + \text{APV}(i) + \text{APV}(r),$$

with the notation for the actuarial present value introduced above.

The computations of all these actuarial present values in the discrete setup are illustrated in Example 3.2.6. In all these expressions we need the annuity values. We use the same annuity function for computing all the three actuarial present values, APV(r), APV(w), and APV(i). In practice, a suitable life table may be used to compute the annuity values separately for each case. It is to be noted that the annuity function involved in APV(r), APV(w), and APV(i) is $\ddot{a}_x^{(12)\eta}$ for $x = \alpha$ to β. After age α and up to β, death and age-service retirement are two causes of decrement, and after age β, death is the only cause of decrement. After retirement, either at age α or β, the pension is payable till the individuals survive. Thus a single decrement model governing the future life time is a suitable model to calculate the annuity function. In the following example, in the calculation of annuity function we assume that the mortality pattern of all the lives after 55 is governed by the single decrement Gompertz model. We find the annuity values under the Gompertz' law and use these in Examples 3.2.7 and 3.2.8 for calculation of actuarial present values of the retirement benefits. The force of interest to calculate the annuity values may not be the same as that in the calculation of the actuarial present value of the benefits.

Example 3.2.6 Suppose that the mortality pattern after age 55 is modeled by the Gompertz law defined by the force of mortality $\mu_x = BC^x$. Suppose $B = 0.00011$ and $C = 1.095$. Find $\ddot{a}_x^{(12)\eta}$ for $x = 55, 56, \ldots, 65$, taking the force of interest $\delta = 0.05$ and rate η of annual increase in pension payment to be $\eta = 0.045$.

Solution It has already been proved that the annuity function $\ddot{a}_y^{(12)\eta}$ with the force of interest δ is the same as $\ddot{a}_y^{(12)}$ with the force of interest $\delta' = \delta - \eta$. Thus, to incorporate the annual increase in pension payment, we have to only modify the force of interest in the annuity function. Further, we know that the monthly annuity function $\ddot{a}_x^{(12)}$ is related to $\ddot{a}_x$ as

$$\ddot{a}_x^{(12)} = \alpha(12)\ddot{a}_x - \beta(12), \quad \text{where } \alpha(12) = \frac{id}{i^{(12)}d^{(12)}} \text{ and } \beta(12) = \frac{i - i^{(12)}}{i^{(12)}d^{(12)}}.$$

The nominal interest rate $i^{(12)}$ and nominal discount rate $d^{(12)}$ are given by

$$i^{(12)} = 12\{(1+i)^{1/12} - 1\} \quad \text{and} \quad d^{(12)} = 12(1 - v^{1/12}).$$

By definition, $\ddot{a}_x = \sum_{k=0}^{\infty} v^k \,_k p_x$. For the Gompertz law with force of mortality $\mu_x = BC^x$, the survival function $S(x)$ is given by $S(x) = e^{-m(C^x - 1)}$, where $m = B/\log_e C$. Hence,

$$_k p_x = \frac{S(x+k)}{S(x)} = \frac{e^{-m(C^{x+k} - 1)}}{e^{-m(C^x - 1)}} = e^{mC^x - mC^{x+k}}, \quad \text{and}$$

$$\ddot{a}_x = \sum_{k=0}^{\infty} v^k e^{mC^x - mC^{x+k}} = e^{mC^x} \sum_{k=0}^{\infty} v^k e^{-mC^x C^k}.$$

We first find $\ddot{a}_{65}$ using the formula $\ddot{a}_{65} = \sum_{k=0}^{\infty} v^k {}_k p_{65}$ and then use the backward recurrence relation $\ddot{a}_x = 1 + v p_x \ddot{a}_{x+1}$ to find $\ddot{a}_x$ for $x = 55$ to 64. With these values of $\ddot{a}_x$, we find $\ddot{a}_x^{(12)}$ using $\alpha(12)$ and $\beta(12)$. The following is a set of R commands for all these computations:

```
b <- 0.00011    #B;
a <- 1.095   #C;
m <- b/log(a, base=exp(1));
e <- exp(1);
del <- 0.05;
eta <- 0.045;
v <- e^(-(del-eta));
x <- 65;
j <- m*a^x;
k <- 0:(100-x);
ad <- e^j*sum(v^k*e^(-j*a^k))    #ä_65^η;
x <- 55:64;
j <- m*a^x;
p <- e^(j-j*a)   #vector of p_x for x=55 to 64;
ad1 <- c(1:10, ad)   #a dummy vector to store the values
          #of ä_x for x=55 to 65, last element, being ä_65^η;
 {
 for (i in 1:10)
 ad1[11-i] <- 1+v*p[11-i]*ad1[11-i+1];
 }
ad1   #a vector of ä_x^η for x=55 to 65;
int <- 1/v-1   #effective rate of interest;
d <- 1-v   #effective rate of discount;
i12 <- 12*((1+int)^(1/12)-1)    #monthly nominal rate
                                #of interest;
d12 <-12*(1-v^(1/12))  #monthly nominal rate of discount;
b1 <- int*d/(i12*d12)    #α(12);
b2 <- (int-i12)/(i12*d12)   #β(12);
ad2 <- round((b1*ad1-b2),4)  #a vector of ä_x^(12)η for x=55
                             #to 65, rounded to 4 decimals;
x <- 55:65;
d1 <- data.frame(x, ad2);
d1   #Table 3.8;
```

Table 3.8 displays the values of $\ddot{a}_x^{(12)\eta}$ for $x = 55$ to 65. From the table we see that the $\ddot{a}_x^{(12)\eta}$ value decreases as age increases, as expected.

Suppose that the initial monthly pension for an individual retiring at age 65 is Rs 10000/-; then the actuarial present value at age 65 of the future pension payments, which increase at the rate of e^{η} per annum, is $12 \times 10000 \times \ddot{a}_{65}^{(12)\eta} = 1267788.00$. This means that at age 65, amount 1267788.00 has to be deposited in the pension

Table 3.8 Discrete monthly life annuity

Age x	$\ddot{a}_x^{(12)\eta}$	Age x	$\ddot{a}_x^{(12)\eta}$
55	16.4839	61	12.7786
56	15.8376	62	12.2045
57	15.2022	63	11.6439
58	14.5781	64	11.0973
59	13.9659	65	10.5649
60	13.3659		

fund so that the fund will raise at the force of interest δ to pay for the monthly pension for the individual at the stated rate.

In the following example we obtain the actuarial present value of the benefit corresponding to a hypothetical pension plan. We use the annuity values as obtained in Example 3.2.6 to find the actuarial present value of the benefit.

Example 3.2.7 Suppose that the retirement benefit plan for the individual who enters at age 25 and who is in active service at age 40 is as follows.

 (i) The age-service retirement benefit is the monthly life annuity due at an initial annual rate of 0.7 % of the average salary of last five years multiplied by the total years of service, with annual increase at the rate of e^η with $\eta = 0.045$. Suppose $\alpha = 55$ and $\beta = 65$.
 (ii) If he withdraws after at least 20 years of service and before age 55, the withdrawal benefit is the deferred life annuity at an annual rate of 0.7 % of the average salary of last five years multiplied by the total years of service, with annual increase at the rate of e^η with $\eta = 0.045$.
(iii) If he retires due to disability before age 55, the disability benefit is the deferred life annuity as in the case of withdrawal.
(iv) If death occurs before 65, then the death benefit is thrice the salary at the age of death, paid at the end of year of death.

Suppose that the annual salary of the individual at age 40 is Rs 500000/- and the decrement pattern is as specified in multiple decrement table, Table 3.4. Find the actuarial present value at age 40 of the benefit to the individual. Use the annuity values for all the three cases as obtained in Example 3.2.6 and the force of interest $\delta = 0.05$.

Solution We use the formulas for the actuarial present values for the given plan as obtained for the discrete setup. In case of age-service retirement, the initial retirement benefit is $R_3(25, 15, k)$ with $d = 0.007$. Hence,

$$\text{APV}(r) = \sum_{k=15}^{25} v^k \, {}_k p_{40}^{(\tau)} \, q_{40+k}^{(r)} R_3(25, 15, k) \ddot{a}_{40+k}^{(12)\eta}.$$

If he withdraws after at least 20 years of service and before age 55, the withdrawal benefit is the deferred life annuity at an annual initial rate of $R_3(25, 15, k)$ with $d = 0.007$. Hence,

$$\text{APV}(w) = \sum_{k=5}^{14} v^k \, _k p_{40}^{(\tau)} q_{40+k}^{(w)} R_3(25, 15, k) \, _{(15-k)|} \ddot{a}_{40+k}^{(12)\eta}$$

$$= v^{15} \ddot{a}_{55}^{(12)\eta} \, _{15} p_{40}^{(\tau)} \sum_{k=5}^{14} q_{40+k}^{(w)} R_3(25, 15, k).$$

If he retires due to disability before age 55, the disability benefit is the deferred life annuity at an initial annual rate of $R_3(25, 15, k)$ with $d = 0.007$. Hence,

$$\text{APV}(i) = \sum_{k=0}^{14} v^k \, _k p_{40}^{(\tau)} q_{40+k}^{(i)} R_3(25, 15, k) \, _{(15-k)|} \ddot{a}_{40+k}^{(12)\eta}$$

$$= v^{15} \ddot{a}_{55}^{(12)\eta} \, _{15} p_{40}^{(\tau)} \sum_{k=0}^{14} q_{40+k}^{(i)} R_3(25, 15, k).$$

The actuarial present value of death benefit is given by

$$\text{APV}(d) = \sum_{k=0}^{24} v^{k+1} \, _k p_{40}^{(\tau)} q_{40+k}^{(d)} 3(ES)_{40+k}.$$

The actuarial present value APV(40) of all the benefits of the plan at age 40 for (25) is then obtained as

$$\text{APV}(40) = \text{APV}(r) + \text{APV}(w) + \text{APV}(i) + \text{APV}(d).$$

The decrement and survival probabilities for each age are corresponding to the decrement model as specified in Table 3.4. Suppose that two columns, the age and salary scale function, of the service table as given in Table 3.3 are stored on D drive as tab delimited Excel file `service.txt`. Suppose that annuity values $\ddot{a}_x^{(12)\eta}$ for $x = 55$ to 65, as obtained in Example 3.2.6, are stored as a text file `ad.txt`, second column specifying the values of annuity for various ages. Further, we use the decrement and the survival probabilities as given in Table 3.4 which is stored as a text file, `mdt.txt`, on drive D. We compute the $(ES)_x$ and $R_3(a, h, k)$ values using R commands similar to those in Example 3.2.5. The following R commands compute all these actuarial present values:

```
del <- 0.05  # force of interest;
e <- exp(1);
v <- e^(-del)  # effective rate of discount;
ad <- read.table("D://ad.txt")  # ä_x^(12) for x = 55 to 65;
```

```
z <- read.table("D://service.txt")
                               # columns 1 and 7 of Table 3.3;
u <- z[16:40, 2]   # w_x values for x = 40 to 64;
v1 <- u/z[16, 2]   # w_{40+k}/w_{40} for k = 0 to 24;
v2 <- 500000*v1   # (ES)_{40+k} from k = 0 to 24;
x <- z[11:40, 2]   # w_x values for x = 35 to 64;
y <- 1:26   # a dummy vector to store the values of
            # 5-year average salary rate;
for (i in 1:26)
  {
  y[i] <- (x[i]+x[i+1]+x[i+2]+x[i+3]+x[i+4])/5
  }
y   # a vector of the values of 5-year averages;
k <- 0:25;
y2 <- (15+k)*.007*500000*y/z[16, 2]
                               # R_3(25,15,k) for k = 0 to 25;
u <- read.table("D://mdt.txt", header=T)   # Table 3.4;
u2 <- u[, 2]   # q_x^{(d)} values for x = 25 to 65;
u3 <- u[, 3]   # q_x^{(w)} values for x = 25 to 65;
u4 <- u[, 4]   # q_x^{(i)} values for x = 25 to 65;
u5 <- u[, 5]   # q_x^{(r)} values for x = 25 to 65;
u6 <- u[, 6]   # p_x^{(\tau)} values for x = 25 to 65;
u1 <- u6[16:40]   # p_x^{(\tau)} values for x = 40 to 64;
p <- c(1, 2:26)   # a dummy vector to store the values of
                  # _k p_{40}^{(\tau)} for k = 0 to 25;
for (i in 2:26)
  {
  p[i] <- p[i-1]*u1[i-1]
  }
p   # a vector of _k p_{40}^{(\tau)} for k = 0 to 25;
v3 <- v^k;
apvr <- sum(v3[16:26]*p[16:26]*u5[31:41]*y2[15:25]
         *ad[, 2]);
apvw <- p[16]*ad[1, 2]*v^(15)*sum(u3[21:30]*y2[6:15]);
apvi <- p[16]*ad[1, 2]*v^(15)*sum(u4[16:30]*y2[1:15]);
apvd <- 3*sum(v3[2:26]*p[1:25]*u2[16:40]*v2[1:25]);
apvr; apvw; apvi; apvd;
apv <- apvr+apvw+apvi+apvd;
apv   # APV(40);
```

We get $APV(r) = 888536.10$, $APV(w) = 48997.45$, $APV(i) = 38180.51$, $APV(d) = 198495.00$, and $APV(40) = 1174209.00$. Thus, the actuarial present value at age 40 of the benefits specified in the pension plan is 1174209.00 for the individual who becomes member of the plan at age 25. Thus, for this individual at his age 40, say at time t, the amount 1174209.00 has to be deposited in the pension fund to pay for the benefits as specified in the plan.

Suppose that there are n_{40} individuals of age 40 at time t, who started their service at age 25. Then $n_{40}\text{APV}(40)$ is the actuarial present value at age 40 of the benefits to such a group of individuals. In general, if at time point t there are n_x individuals of age x, $a \leq x \leq \beta$, then the actuarial present value of the benefits corresponding to all the individuals in the plan at time t can be obtained as the sum of the actuarial present values corresponding to all the individuals of various ages at time t. Such actuarial present value will indicate the amount of pension cost at time t and hence the amount of deposit in the pension fund, which will raise at a force of interest δ, to be sufficient to provide pension benefits to the individuals whenever they are eligible.

The pension fund is generated by annual contributions to the fund. The annual contribution for all employees together is the annual contribution to the pension fund. The contribution may be by the employee or employer or by both in some defined proportion. Pension benefits are financed by payments connected to an individual's salary levels and continue up to retirement or the maximum retirement age. There are two basic plans for contributions. The first is a flat rate, and the second method is a flat percentage of the individual's salary for designated years.

The following example illustrates the computation of rate of contribution, proportional to the salary rate, for an employee of age 25 when he enters the service, again in a hypothetical setup. In the next chapter we will discuss how to find a pension fund for a group of individuals at a specified time t.

Example 3.2.8 Suppose that the retirement benefit plan for the individual who enters at age 25 is as follows.

(i) The age-service retirement benefit is the monthly life annuity due at an annual initial rate of 0.7 % of the average salary of last five years multiplied by the total years of service, that is, $R_3(25, 0, k)$. Further, assume that annual increase in the annual pension benefit is at the rate $\eta = 0.045$. Suppose $\alpha = 55$ and $\beta = 65$.

(ii) If he withdraws after at least 20 years of service and before age 55, the withdrawal benefit is the deferred life annuity at the same rate as in (i).

(iii) If he retires due to disability before age 55, the disability benefit is the deferred life annuity at the same rate as in (i). No benefit will be paid if disability occurs within first five years.

(iv) If death occurs before 65, then the death benefit is thrice the salary at the age of death and is paid at the end of year of death.

The annual contribution is proportionate to the annual salary. Suppose that the annual salary of the individual at age 25 is Rs 240000/- and the salary scale is as in the last column of Table 3.3. Suppose that the decrement pattern is as specified in the multiple decrement table, Table 3.4. The annuity values for all the three cases are according to the Gompertz model with $B = 0.00011$ and $C = 1.095$ and the force of interest $\delta = 0.05$.

Find the rate of annual contribution to the pension fund for (25). Prepare a table of projected salary and projected year wise contribution to the pension fund for (25).

Solution We first find the actuarial present value at age 25 of the benefits as specified in the plan. To find it, we adopt the similar procedure as in Example 3.2.7. The retirement benefit is the monthly life annuity due at an initial annual rate of 0.7 % of the average salary of last five years multiplied by the total years of service, that is, $R_3(25, 0, k)$ with $d = 0.007$. The minimum age for age-service retirement is 55, and 65 is the age of mandatory retirement. Hence, the actuarial present value APV(r), at age 25, of the retirement benefits is given by

$$\text{APV}(r) = \sum_{k=30}^{40} v^k \, {}_k p_{25}^{(\tau)} q_{25+k}^{(r)} R_3(25, 0, k-1) \ddot{a}_{25+k}^{(12)\eta}.$$

If he withdraws after at least 20 years of service and before age 55, the withdrawal benefit is the deferred life annuity at an initial annual rate of $R_3(25, 0, k)$. Hence,

$$\text{APV}(w) = \sum_{k=20}^{29} v^k \, {}_k p_{25}^{(\tau)} q_{25+k}^{(w)} R_3(25, 0, k) \, {}_{(30-k)|}\ddot{a}_{25+k}^{(12)\eta}$$

$$= v^{30} \ddot{a}_{55}^{(12)\eta} \, {}_{30} p_{25}^{(\tau)} \sum_{k=20}^{29} q_{25+k}^{(w)} R_3(25, 0, k).$$

If he retires due to disability before age 55, the disability benefit is the deferred life annuity at an initial annual rate of $R_3(25, 0, k)$. Hence,

$$\text{APV}(i) = \sum_{k=5}^{29} v^k \, {}_k p_{25}^{(\tau)} q_{25+k}^{(i)} R_3(25, 0, k) \, {}_{(30-k)|}\ddot{a}_{25+k}^{(12)\eta}$$

$$= v^{30} \ddot{a}_{55}^{(12)\eta} \, {}_{30} p_{25}^{(\tau)} \sum_{k=5}^{29} q_{25+k}^{(i)} R_3(25, 0, k).$$

The actuarial present value of death benefit is given by

$$\text{APV}(d) = \sum_{k=0}^{39} v^{k+1} \, {}_k p_{25}^{(\tau)} q_{25+k}^{(d)} 3(ES)_{25+k}.$$

The actuarial present value APV(25) of all the benefits of the plan at age 25 for (25) is then obtained as

$$\text{APV}(25) = \text{APV}(r) + \text{APV}(w) + \text{APV}(i) + \text{APV}(d).$$

The second step is to find the actuarial present value of the contributions, defined as the proportion of the annual salary paid for the employee till he is in active service. Suppose that the contributions are paid at the end of each year. The annual contribution at the end of year $25 + u$ is $P \frac{w_{25+u-1}}{w_{25}}$, $u = 1, 2, \ldots, 40$, where P is a fraction. There are four modes of decrement from the active service. We begin with the decrement due to age-service retirement. Suppose that (25) is in active service till 55 and may retire between 55 to 65 or will complete 40 years of service and retires at age 65. When (25) completes one year of service, a fraction P of his salary at 25, w_{25}, will be a contribution to the pension fund at the end of the year; its present value at 25 is then obtained by multiplying by v. Thus, when (25) completes k years, with probability $_kp_{25}$, $k = 1, 2, \ldots, 30$, $P\frac{w_{25+k-1}}{w_{25}}$ will be contributed to the pension fund at the end of kth year, its present value at age 25 is obtained by multiplying by v^k. If (25) works till 55 and retires between 55 to 56 with probability $_{30}p_{25}^{(\tau)}q_{55}^{(r)}$, then the contribution is $P\frac{w_{55}}{w_{25}}$ will be contributed to the pension fund at the end of the year; its present value at age 25 is obtained by multiplying by v^{31}. If (25) works till 56, and retires between 56 to 57 with probability $_{31}p^{(\tau)}{}_{25}q_{56}^{(r)}$, then the present value of the contribution is $P\frac{w_{55}v^{30}+w_{56}v^{31}}{w_{25}}$. Continuing on similar lines, if (25) works till 55 and retires between $55 + k$ to $55 + k + 1$ for $k = 0, 1, \ldots, 9$, then the present value of contribution to pension fund is $[\sum_{u=31}^{31+k} v^u \frac{w_{25+u-1}}{w_{25}}]P$. If he is in service till 65, then the present value of the contribution to the pension fund will be $P\frac{w_{64}}{w_{25}}v^{40}$. Thus, the actuarial present value A_1 of annual contribution, in the case of age-service retirement, is given by

$$
A_1 = P\left\{ \sum_{k=1}^{30} {}_kp_{25}^{(\tau)} v^k \frac{w_{25+k-1}}{w_{25}} + \sum_{k=30}^{39} {}_kp_{25}^{(\tau)} q_{25+k}^{(r)} \left[\sum_{u=31}^{k+1} v^u \frac{w_{25+u-1}}{w_{25}} \right] \right.
$$

$$
\left. + {}_{40}p_{25}^{(\tau)} v^{40} \frac{w_{25+39}}{w_{25}} \right\}.
$$

If he exits the service either due to withdrawal or disability, then the contributions are made till he is in active service. In this case the actuarial present value A_2, at age (25), of the contributions is given by

$$
A_2 = P\left\{ \sum_{k=0}^{29} {}_kp_{25}^{(\tau)} \left(q_{25+k}^{(i)} + q_{25+k}^{(w)} \right) \left[\sum_{u=1}^{k+1} v^u \frac{w_{25+u-1}}{w_{25}} \right] \right\}.
$$

If he exits the service due to death, then the contributions are made till he is in active service. In this case the actuarial present value A_3, at age (25), of the contributions is given by

$$
A_3 = P\left\{ \sum_{k=0}^{39} {}_kp_{25}^{(\tau)} q_{25+k}^{(d)} \left[\sum_{u=1}^{k+1} v^u \frac{w_{25+u-1}}{w_{25}} \right] \right\}.
$$

It is to be noted that $q_{25+k}^{(w)}$ and $q_{25+k}^{(i)}$ is 0 after $25 + k = 55$. Hence A_2 and A_3 can be combined. Thus the actuarial present value at age 25 of contributions in case of exit due to death or withdrawal or disability, denoted A_4, is given by

$$
A_4 = P\left\{ \sum_{k=0}^{39} {}_kp_{25}^{(\tau)} \left(q_{25+k}^{(d)} + q_{25+k}^{(i)} + q_{25+k}^{(w)} \right) \left[\sum_{u=1}^{k+1} v^u \frac{w_{25+u-1}}{w_{25}} \right] \right\}.
$$

Combining all the modes of decrement, the actuarial present value of the contributions, APV(C), at age 25 is obtained. It is to be noted that

$$APV(C) = (AS)_{25} \sum_{j=1}^{4} E[Y_j | J = j] P[J = j],$$

where $E[Y_j | J = j]$ is the actuarial present value of contributions if (25) exits due to mode j. We have $E[Y_1 | J = 1]P[J = 1] + E[Y_2 | J = 2]P[J = 2] + E[Y_3 | J = 3]P[J = 3] = A_4$ and $E[Y_4 | J = 4]P[J = 4] = A_1$. Hence, APV($C$) is given by

$$APV(C) = (AS)_{25}\{A_1 + A_4\},$$

where $(AS)_{25} = 240000$. We find P by equating the actuarial present value of the benefits to the actuarial present value of the contributions. The projected annual contribution is then obtained as proportion P of projected annual salary at various ages with initial salary at age 25 to be Rs 240000/-. The following set of R commands is used for these computations; the R commands to compute the actuarial present value of the benefits are similar to those in Example 3.2.7:

```
del <- 0.05;
e <- exp(1);
v <- e^(-del);
z <- read.table("D://service.txt", header=T)
                                    #Table of w_x values;
u <- z[, 2]   #w_x values for x = 25 to 64;
v1 <- u/z[1, 2]   # w_{25+k}/w_{25} for k = 0 to 39;
as25 <- 240000   #annual salary at age 25;
v2 <- as25*v1   # (ES)_{25+k} = (AS)_{25} w_{25+k}/w_{25} for k = 0 to 39;
y <- 1:36   #dummy vector to store _5Z_{25+k} for k = 5 to 40;
for (i in 1:36)
  {
  y[i] <- (u[i]+u[i+1]+u[i+2]+u[i+3]+u[i+4])/5
  }
y   #vector of _5Z_{25+k} for k = 5 to 40;
k <- 5:40;
y2 <- (k)*.007*as25*y/z[1, 2]   # R_3(25, 0, k);
u <- read.table("D://mdt.txt", header=T)   #Table 3.4;
u3 <- u[, 3]   #q_x^{(w)};
u4 <- u[, 4]   #q_x^{(i)};
u6 <- u[, 6]   #p_x^{(τ)};
p <- c(1, 2:41) #dummy vector to store _kp_{25}^{(τ)} for k = 0 to 40;
for (i in 2:41)
  {
  p[i] <- p[i-1]*u6[i-1]
  }
```

```
p   #vector of kp25^(τ) for k=0 to 40;
u1 <- read.table("D://jointjk.txt", header=T)  #Table 3.5;
qd <- u1[, 2]   # kp25^(τ) qx+k^(d);
qw <- u1[, 3]   # kp25^(τ) qx+k^(w);
qi <- u1[, 4]   # kp25^(τ) qx+k^(i);
qr <- u1[, 5]   # kp25^(τ) qx+k^(r);
q  <- u1[, 6]   # kp25^(τ) qx+k^(τ);
k  <- 0:40;
v3 <- v^k;
ad <- read.table("D://ad.txt")
                           #Table 3.8 of annuity values;
apvr <- sum(v3[31:41]*qr[31:41]*y2[25:35]*ad[, 2]);
apvw <- p[31]*ad[1,2]*v^(30)*sum(u3[21:30]*y2[16:25]);
apvi <- p[31]*ad[1,2]*v^(30)*sum(u4[6:30]* y2[1:25]);
apvd <- 3*sum(v3[2:41]*qd[1:40]*v2[1:40]);
apvr;  apvw;  apvi;  apvd;
apv <- apvr+apvw+apvi+apvd;
apv;
s1 <- cumsum(v3[32:41]*v1[31:40]);
s2 <- sum(p[2:31]*v3[2:31]*v1[1:30])+sum(qr[31:40]*s1)
       +p[41]*v3[41]*v1[40];
s3 <- cumsum(v3[2:41]*v1[1:40]);
s4 <- sum(p[1:40]*(qd[1:40]+qw[1:40]+qi[1:40])*s3);
s5 <- as25*(s2+s4)   #APV(C);
s5;
cr <- apv/s5;
cr  #P;
con <- v2*cr;
age <- 25:65;
d <- data.frame(age, v2, con);
d   #Table 3.9;
```

For the given data, we get, $APV(r) = 158887.10$, $APV(w) = 8761.53$, $APV(i) = 7130.42$, $APV(d) = 39048.68$, $APV(25) = 213827.80$, and $APV(C) = 3461488.00$. The proportion $P = 0.061773$, that is, 6.18 % of the annual salary will be the annual contribution to the pension fund. Table 3.9 displays the projected annual salary $(ES)_x$ for age $x = 25$ to 64 and the projected annual contribution C_{x+1} to the pension fund for age $x = 25$ to 64.

The discussion so far pertains to the actuarial present values on an individual basis that is cost of the plan per individual. An employer requires the knowledge of the cost of the plan on an aggregate basis for the whole group of employees working with the employer. Aggregate values can be obtained by summing over the values on individual basis. The cost of the pension plan at a specific time is thus the actuarial present value of the benefits to all the employees, as assured in the pension plan.

Table 3.9 Predicted annual salary and contribution

Age x	$(ES)_x$	C_{x+1}	Age x	$(ES)_x$	C_{x+1}
25	240000.00	14825.61	45	676800.00	41808.21
26	252000.00	15566.89	46	712800.00	44032.05
27	266400.00	16456.42	47	751200.00	46404.15
28	280800.00	17345.96	48	792000.00	48924.50
29	295200.00	18235.49	49	830400.00	51296.60
30	312000.00	19273.29	50	876000.00	54113.46
31	326400.00	20162.82	51	921600.00	56930.33
32	345600.00	21348.87	52	967200.00	59747.19
33	362400.00	22386.66	53	1017600.00	62860.57
34	384000.00	23720.97	54	1068000.00	65973.94
35	403200.00	24907.02	55	1120800.00	69235.58
36	424800.00	26241.32	56	1173600.00	72497.21
37	448800.00	27723.88	57	1228800.00	75907.10
38	470400.00	29058.19	58	1284000.00	79316.99
39	496800.00	30689.00	59	1339200.00	82726.88
40	523200.00	32319.82	60	1392000.00	85988.51
41	549600.00	33950.64	61	1447200.00	89398.40
42	580800.00	35877.97	62	1504800.00	92956.55
43	612000.00	37805.29	63	1564800.00	96662.95
44	643200.00	39732.62	64	1629600.00	100665.86

There are many funding or budgeting methods available to assure that contributions are made to the plan in an orderly and appropriate manner. Next chapter discusses various funding methods to meet the cost of the plan. In the following we discuss briefly other type of pension plan, that is, defined contribution pension plan.

Defined Contribution Plan In a defined contribution plan, contributions are paid into an individual account for each member. The contributions are invested, for example, in the stock market, and the returns on the investment (which may be positive or negative) are credited to the individual's account. On retirement, the member's account is used to provide retirement benefits, sometimes through the purchase of an annuity which then provides a regular income. Alternatively, an accumulated amount can be made available to the individual at the time of retirement, as lump some amount. In this plan the actuarial present value is simply the accumulation, under specified rate of interest, of the contributions made by the individual or by the employer for the individual. Thus, such an accumulation generates a fund from which benefit, usually a monthly annuity, is given to the individual. The defined contribution rate is determined with a retirement income as a goal. In defined contribution pension plan the benefit on exit is the same, irrespective of the mode of exit.

Defined contribution plans have become widespread all over the world in recent years and are now the dominant form of plan in the private sector in many countries. For example, the number of defined benefit plans in the US has been steadily declining, as more and more employers see pension contributions as a large expense avoidable by terminating the defined benefit plan and instead offering a defined contribution plan. Money contributed can either be from employee salary or from employer contributions. In a defined contribution plan, investment risk and investment rewards are assumed by each individual/employee/retiree and not by the sponsor/employer. The cost of a defined contribution plan is readily calculated, but the benefit from a defined contribution plan depends upon the account balance at the time an employee wishes to use the assets. So, for this arrangement, the contribution is known, but the benefit is unknown (until calculated). Despite the fact that the participant in a defined contribution plan typically has control over investment decisions, the plan sponsor retains a significant degree of fiduciary responsibility over investment of plan assets, including the selection of investment options and administrative providers.

Assumptions about Expenses and Investment Returns   The expenses of administering the pension plan must be added to the benefits paid in arriving at the cost of the plan. The expense assumptions depend on the type of administration and funding instrument involved. Under individual policy plans and some group pension contracts, the insurance company includes a loading factor for expenses in gross premiums charged for benefits. The investment income earned on the accumulated assets of a funded pension plan reduces the ultimate cost of the plan. The choice of the appropriate rate of investment return is particularly difficult if a sizable portion of the assets is invested in common stocks, since these investments are subject to significant fluctuations in value. The investment return is the most important factor affecting the cost of the plan. The investment income earned on the accumulated assets of a funded plan reduces the ultimate cost of the plan.

In Chap. 4 we study various methods of pension funding.

3.3 Exercises

3.1 Suppose that a member of the pension plan, joining the plan at age 25, exits the plan by any one of the four causes of exit: withdrawal, disability retirement, death in service, and age retirement. The probability law governing these decrements is as specified in Example 3.2.1. Calculate the probability that a member of age 35
  (i) retires at exact age 65;
  (ii) exits the plan due to death in service or due to withdrawal or due to disability or due to age-service retirement.
3.2 Suppose that the annual salary of an employee aged exactly 35 is Rs 500000/- and salary increases according to the salary scale function given in Table 3.3.

(i) Estimate the employee's salary for next 10 years.
(ii) Suppose that the pensionable salary for the pension benefit is defined as the average salary in three years before retirement. Calculate the pensionable salary if retirement is at age 65.
(iii) Suppose that an individual is eligible for the retirement benefit if he retires at complete age between 55 to 65. Assume that $d = 0.19$ if the initial pension benefit rate is fraction d of the final salary rate and fraction d of the 3-year average salary rate. Suppose that $d = 0.008$ if the benefit rate is fraction d of the 3-year average salary rate multiplied by the total years of service. Find at age 35 the projected annual initial pension benefit for ages 55 to 65.

3.3 Suppose that the retirement benefit plan for the individual who enters at age 25 and who is in active service at age 35 is as follows.
(i) The age-service retirement benefit is the monthly life annuity due at an initial annual rate of 0.7 % of the average salary of last five years multiplied by the total years of service, with annual increase at the rate of e^η with $\eta = 0.045$. Suppose $\alpha = 55$ and $\beta = 65$.
(ii) If he withdraws after at least 10 years of service and before age 55, the withdrawal benefit is the deferred life annuity at an annual rate of 0.7 % of the average salary of last five years multiplied by the total years of service, with annual increase at the rate of e^η with $\eta = 0.045$.
(iii) If he retires due to disability before age 55, the disability benefit is the deferred life annuity as in the case of withdrawal.
(iv) If death occurs before 65, then the death benefit is thrice the salary at the age of death, paid at the moment of death.
Suppose that the annual salary of the individual at age 35 is Rs 500000/- and the decrement pattern is as specified in Example 3.2.1. Find the actuarial present value at age 35 of the benefit to the individual. The annuity values for all the three cases are according to the Gompertz model with $B = 0.00011$ and $C = 1.095$ and the force of interest $\delta = 0.05$.

3.4 Repeat Exercise 3.3 if the age-service retirement benefit is paid in advance and are guaranteed for five years.

3.5 Suppose that employees in a defined contribution pension plan pay contribution of 6 % of their annual salary at the end of each year. Calculate the APV at entry, of contributions for a new entrant at age 35 with starting annual salary of Rs 500000. Suppose that the effective rate of interest is 6 % and salary scale is as given in Table 3.3. Use the multiple decrement model as specified in Table 3.3.

3.6 Suppose that the retirement benefit plan for the individual who enters at age 25 is as follows.
(i) The age-service retirement benefit is the monthly life annuity due at an annual initial rate of 0.7 % of the average salary of last five years multiplied by the total years of service, that is, $R_3(25, 0, k)$. Further, assume that the annual increase in the annual pension benefit is at the rate $\eta = 0.045$. Suppose $\alpha = 55$ and $\beta = 65$.
(ii) If he withdraws after at least 15 years of service and before age 55, the withdrawal benefit is the deferred life annuity at the same rate as in (i).

(iii) If he retires due to disability before age 55, the disability benefit is the deferred life annuity at the same rate as in (i). No benefit will be paid if disability occurs within first five years.

(iv) If death occurs before 65, then the death benefit is thrice the salary at the age of death and is paid at the end of year of death.

The annual contribution is in proportion of the annual salary. Suppose that the annual salary of the individual at age 25 is Rs 300000/- and the salary scale is as in the last column of Table 3.3. Suppose that the decrement pattern is as specified in Table 3.3. The annuity values for all the three cases are according to the Gompertz model with $B = 0.00009$ and $C = 1.096$ and the force of interest $\delta = 0.06$.

Find the rate of annual contribution to the pension fund by (25). Prepare a table of projected salary and projected year wise contribution to the pension fund for (25).

Chapter 4
Pension Funding

4.1 Introduction

A pension plan is a system for purchasing deferred life annuities, payable during retirement, and certain ancillary benefits, with a temporary annuity of contributions during active service. In a typical defined benefit pension plan the cost of an employee benefit is funded by the employer. Employer's contribution is valued regularly and is usually expressed as a percentage of the salary. With an insurance policy, the policyholder pays for a contract through a level periodic premiums or a single premium. However, in pension plan the level of contribution from the employer is not a part of the contract, and it need not be a level. The contributions need to be adjusted from time to time. The main principle in deciding the rate of contribution is as follows. The funding level for the year is set so that the amount required to be paid, together with the fund value at the start of the year, is sufficient to pay the expected cost of any benefits due during the year and to pay the expected cost of establishing the new actuarial liability at the year end. We discuss these issues in detail in this chapter.

Cost of the pension plan at a specific time is the actuarial present value of the benefits to the employees, as assured in the pension plan. In Chap. 3, we have discussed various factors affecting the ultimate cost of the plan. Once the estimate of the ultimate cost of the plan is determined, the next step is to determine the contributions required to pay for the estimated cost in an orderly manner, so that the estimated cost of the plan is spread over future years. These actuarial techniques are referred to as actuarial cost methods or actuarial funding methods. In Example 3.2.8 in Chap. 3, we have discussed the computation of contribution on individual basis. In this chapter, we consider the individual set up with a different approach and also the aggregate setup.

A funding method specifies the pattern, that is, the frequency, and the amount of aggregate contributions required to balance the benefit payments. Once a funding method is chosen, contributions are generated, which in turn lead to the development of the assets of the pension plan. Ideally assets are accumulated to equal the reserve

S. Deshmukh, *Multiple Decrement Models in Insurance*,
DOI 10.1007/978-81-322-0659-0_4, © Springer India 2012

required at normal retirement date. Thus, the funding plan is a budgeting plan for accumulating the funds necessary to provide the annuity benefits.

The time value of money is an important consideration in pension funding as the outflow of the benefit payouts over time is subject to life and other contingencies, and inflow of contributions occurs at different time points and in a different pattern. Actuary then has to make certain assumptions about the rate of return on the pension fund, age of retirement, rates of turnover and retirement, etc. The most important step is to assign to each fiscal year a portion of the present value of future benefit payments in such a way that the costs are accrued over the working lifetimes of the employees. Any scheme for making such an assignment of costs is the actuarial cost or actuarial funding method. The application of cost method to a particular plan in order to compute its cost is called an actuarial valuation. The primary purpose of the pension valuation is to determine the annual cost of the plan. In the following sections, we discuss various actuarial funding methods. An actuarial cost method establishes the amounts and incidences of the contributions to the pension fund. These contributions are called normal costs or normal contributions. In north America it is commonly known as normal costs. There are supplemental costs pertaining to the benefit and expenses of a pension plan. We will not discuss these in this chapter. There is a wide variety of actuarial funding methods. We discuss some of these below.

4.2 Accrued Benefit Cost Method for an Individual

Accrued benefit cost methods are also known as the unit credit methods. In these methods, the employer (and the employee under contributory plan) sets aside funds on some systematic basis prior to the employee's retirement date. Thus periodic contributions are made on behalf of the group of active employees during their working years. In this approach, for cost determination purpose, an employee's benefits under the pension plan are deemed to accrue in direct relation to the years of service. In accrued benefit funding methods, the normal costs are based directly upon benefits accrued to the date of cost determination. These focus on maintaining a certain level of funding and are security driven, in that they attempt to establish and maintain a sound relationship between the fund assets and the accrued liabilities. The funding requirement is then the contributions required to achieve the funding objective. All these methods differ by a rate at which prospective pension obligations are recognized during the participant's working lifetimes.

The first step in the calculation of normal cost in the accrued benefit cost method is to determine the present value of each participant's pension for the year for which costs are being calculated. The second step is to define an appropriate accrual function. Let r denote the age of mandatory retirement; the provision of voluntary retirement is ignored at this stage. To express the accrual of actuarial liability for a pension commencing at age r, an accrual function $M(x)$ is defined to represent a fraction of the actuarial value of future pensions, expecting that $M(x) \times$ actuarial present value at age x of future benefits equals the liability at age x and will accumulate to a fund sufficient to pay the life annuity of pension benefits. It is assumed that $0 \le M(x) \le 1$, basically to have a common metric for the sake of comparison. Further it is assumed that $M(x)$ is a nondecreasing, right-continuous function of age variable. There is no liability before age a, the age of entry to the pension plan, and hence $M(a) = 0$. For all funding methods requiring accrual or recognition of the total liability by age r, $M(x) = 1$ for $x \ge r$. As an illustration, under initial funding, if entire liability for the future pensions is recognized when the participant enters at age a, then

$$M(x) = 0 \quad \text{for } x < a \quad \text{and} \quad M(x) = 1 \quad \text{for } x \ge a.$$

Such a method is known as initial funding. On the contrary, under terminal funding, entire liability is recognized when the participant retires. Thus, in terminal funding,

$$M(x) = 0 \quad \text{for } a \le x < r \quad \text{and} \quad M(x) = 1 \quad \text{for } x \ge r.$$

If the benefit is accrued uniformly during the period (a, r) of active service, then $M(x) = (x - a)/(r - a)$. The following example illustrates how the accrual function serves the desired purpose.

Example 4.2.1 Suppose that the pension benefit is a continuous annuity at the rate of 1 unit per annum from retirement age r. This benefit is paid for by a temporary life annuity which pays 1 unit per annum in pension fund, starting from entry age a to the retirement age r. Let $M(x)$ be defined as $M(x) = \frac{\bar{a}_{a:\overline{x-a}|}}{\bar{a}_{a:\overline{r-a}|}}$, $a < x < r$. Show that $M(x)_{r-x|}\bar{a}_x$ is equal to the reserve at age x corresponding to a continuous $r - a$ year deferred life annuity issued at age a.

Solution For a continuous $r - a$ year deferred annuity issued at age a, the reserve at intermediate age x is the accumulation of the fund via annual premiums, up to age x, using retrospective approach. It is to be noted that no benefits are paid, so reserve is just the accumulation of premiums. Thus the reserve is given by

$$\overline{V}_x({}_{r-a|}\bar{a}_a) = \overline{P}({}_{r-a|}\bar{a}_a)\overline{S}_{a:\overline{x-a}|}$$

$$\text{where } \overline{P}({}_{r-a|}\bar{a}_a) = \frac{{}_{r-a|}\bar{a}_a}{\bar{a}_{a:\overline{r-a}|}} \text{ and } \overline{S}_{a:\overline{x-a}|} = \frac{\bar{a}_{a:\overline{x-a}|}}{{}_{x-a}E_a}.$$

Further we have ${}_{x-a}E_a = {}_{x-a}p_a v^{x-a}$ and ${}_{n|}\bar{a}_x = {}_n E_x \bar{a}_{x+n}$. Hence the expression for reserve simplifies to

$$\overline{V}_x({}_{r-a|}\bar{a}_a) = \overline{P}({}_{r-a|}\bar{a}_a)\overline{S}_{a:\overline{x-a}|}$$

$$= \frac{{}_{r-a|}\bar{a}_a}{\bar{a}_{a:\overline{r-a}|}} \times \frac{\bar{a}_{a:\overline{x-a}|}}{{}_{x-a}E_a} = \frac{\bar{a}_{a:\overline{x-a}|}}{\bar{a}_{a:\overline{r-a}|}} \times \frac{{}_{r-a}E_a \bar{a}_r}{{}_{x-a}E_a}$$

$$= M(x) \times \frac{r-a p_a v^r \; {}^a \bar{a}_r}{x-a p_a v^{x-a}} = M(x) \times \frac{x-a p_a \; r-x p_x v^{x-a+r-x} \bar{a}_r}{x-a p_a v^{x-a}}$$

$$= M(x) \times {}_{r-x} p_x v^{r-x} \bar{a}_r = M(x) \times {}_{r-x} E_x \bar{a}_r = M(x) \times {}_{r-x} | \bar{a}_x,$$

and the result is proved. Thus the reserve at age x is the accrual function multiplied by the actuarial present value of future benefit.

We interpret the result in Example 4.2.1 as follows. The actuarial present value of unit annual benefit after the retirement till the individual is alive is given by $\bar{a}_r$. Its actuarial present value at age x is ${}_{r-x} E_x \bar{a}_r = {}_{r-x} | \bar{a}_x$. The result shows that $M(x) {}_{r-x} | \bar{a}_x$ represents the reserve at age x or the liability at age x. In other words, the function $M(x) {}_{r-x} | \bar{a}_x$ will accumulate to $\bar{a}_r$ as x increases to r. This is clear from the fact that $M(r) {}_{r-r} | \bar{a}_r = \bar{a}_r$ and

$$\bar{V}_r({}_{r-a} | \bar{a}_a) = \bar{P}({}_{r-a} | \bar{a}_a) \bar{S}_{a:\overline{r-a}|} = \frac{{}_{r-a} | \bar{a}_a}{\bar{a}_{a:\overline{r-a}|}} \times \frac{\bar{a}_{a:\overline{r-a}|}}{{}_{r-a} E_a} = \frac{{}_{r-a} E_a \bar{a}_r}{{}_{r-a} E_a} = \bar{a}_r.$$

Thus the accrual function $M(x)$ presents the part of accrued liability at age x. Hence, it is used to decide the rate of contribution to the fund at age x.

The concept of accrual function is more or less similar to that of distribution function. From the distribution function, whenever it is differentiable, the probability density function is obtained as the derivative of the distribution function. Similarly, to define the rate of change of the function $M(x)$, the pension accrual density function, denoted $m(x)$, is defined as

$$M(x) = \int_a^x m(y) \, dy \quad \Leftrightarrow \quad m(x) = \frac{d}{dx} M(x), \quad x \geq a.$$

Thus if $M(x)$ is analogous to the distribution function, $m(x)$ is the analogue of the probability density function. In general it is assumed that $m(x)$ is continuous for $a < x < r$, right-continuous at a, left continuous at r, and $m(x) = 0$ for $x < a$ and $x > r$. As an illustration, let $M(x) = (x - a)/(r - a)$; then $m(x) = (r - a)^{-1}$. Let $M(x)$ be as defined in Example 4.2.1, that is,

$$M(x) = \frac{\bar{a}_{a:\overline{x-a}|}}{\bar{a}_{a:\overline{r-a}|}} = \frac{\int_a^x e^{-\delta(y-a)} S(y) \, dy}{\int_a^r e^{-\delta(y-a)} S(y) \, dy}.$$

Then,

$$M'(x) = m(x) = \frac{e^{-\delta(x-a)} S(x)}{\int_a^r e^{-\delta(y-a)} S(y) \, dy} = \frac{e^{-\delta x} S(x)}{\int_a^r e^{-\delta y} S(y) \, dy}.$$

With this form of $m(x)$, we have seen that the projected benefit after retirement is funded by a level contribution from entry age to retirement. Further, $m(x)$ is essentially proportional to $S(x) e^{-\delta x}$, and the denominator of $m(x)$ is the norming constant so that $\int_a^r m(x) \, dx = 1$.

The advantage of introducing the accrual function is that we can develop pension theory simultaneously for a whole family of actuarial cost methods rather than separately for each method. Using the accrual function, we now proceed to find the normal cost rate which allocates the actuarial present value of future benefits to the various time points of valuation in a participant's active service. Determination of normal cost rate is the main aim of any funding method. We first discuss how to find the normal cost at individual level and then in the next section extend it for a group.

We use the accrual function $M(x)$ or $m(x)$ to find the normal cost rate and accrued actuarial liability for each member in the plan. Suppose that the age of the individual is x, $a \le x \le r$, and $\bar{a}_r^h$ denotes the actuarial present value of the unit initial benefit rate with increase in the pension benefit for later ages being recognized by the function h as indicated in the superscript h in $\bar{a}_r^h$. Our aim is to find the normal cost at age x, $a < x < r$, so we first find the actuarial present value $(aA)(x)$ of the pension benefit at age x. It is given by

$$(aA)(x) = e^{-\delta(r-x)}\,_{r-x}p_x\bar{a}_r^h = e^{-\delta(r-x)}\frac{S(r)}{S(x)}\bar{a}_r^h = \,_{r-x}E_x\bar{a}_r^h.$$

It is to be noted that the survival function $S(x)$ and the force of interest δ in the definition of $\bar{a}_r^h$ and $(aA)(x)$ may not be the same.

The amount of contribution to the pension fund at age x has to be a fraction of this actuarial present value so that its accumulation equals the liability at age x. This motivates the definition of the normal cost rate $P(x)$ and the accrued actuarial liability $(aV)(x)$ in terms of the accrual function. These are defined as

$$P(x) = (aA)(x)m(x) \quad \text{and} \quad (aV)(x) = (aA)(x)M(x).$$

Thus multiplying $(aA)(x)$ by $M(x)$ gives the accrued liability, while multiplying $(aA)(x)$ by $m(x)$, so-called ordinate at x, gives the amount of contribution at age x. It is to be noted that $(aA)_x$ is the actuarial present value at age x of the unit initial benefit rate with increase in the pension benefit for later ages being recognized by the function h. If the initial annual benefit is b units, then the rate of contribution for each age will be given by $bP(x)$. In other words, if amount $bP(x)$ is deposited in the pension fund from entry age a to the retirement age r, then the fund will be raised at the time of retirement to an amount which will be sufficient to pay the retirement benefit, with initial benefit b units with annual increase as given by the function h.

We illustrate computation of $P(x)$ and $(aV)(x)$ for some accrual functions and with some initial benefit b. We begin with the accrual function $M_1(x)$ introduced in Example 4.2.1. For that $M_1(x)$, the density $m_1(x)$ is given by $m_1(x) = \frac{S(x)e^{-\delta x}}{\int_a^r S(y)e^{-\delta y}\,dy}$. The fact that $m(x)$ is proportional to $S(x)e^{-\delta(x-a)} = \,_{x-a}p_a e^{-\delta(x-a)}$ with $S(a) = 1$ leads to its interpretation as the actuarial present value at entry age a of 1 unit, payable at x, provided that the individual survives to age x.

Example 4.2.2 Let the accrual function be given by

$$m_1(x) = \frac{S(x)e^{-\delta x}}{\int_a^r S(y)e^{-\delta y}\,dy}.$$

Find the normal constant rate $P(x)$ and the accrued actuarial liability $(aV)(x)$.

Solution By definition,

$$P(x) = (aA)(x)m_1(x) = e^{-\delta(r-x)}\frac{S(r)}{S(x)}\bar{a}_r^h \times \frac{S(x)e^{-\delta x}}{\int_a^r S(y)e^{-\delta y}\,dy} = \frac{e^{-\delta r}S(r)\bar{a}_r^h}{\int_a^r S(y)e^{-\delta y}\,dy}.$$

We note that in this case $P(x)$ is free from x, it depends on the force of interest, the actuarial present value of the annuity based on the function h, and the survival function. Further, $(aV)(x)$ is given by

$$(aV)(x) = (aA)(x)M_1(x) = e^{-\delta(r-x)}\frac{S(r)}{S(x)}\bar{a}_r^h \times \frac{\int_a^x S(y)e^{-\delta y}\,dy}{\int_a^r S(y)e^{-\delta y}\,dy} = \overline{V}_x(r-a|\bar{a}_a),$$

as derived in Example 4.2.1.

Remark 4.2.1 Let us rewrite $P(x)$ as

$$P(x) = \frac{e^{-\delta(r-a)}\frac{S(r)}{S(a)}\bar{a}_r^h}{\int_a^r e^{-\delta(y-a)}\frac{S(y)}{S(a)}\,dy}.$$

It is to be noted that the numerator of $P(x)$ is the actuarial present value at entry age a of the retirement benefit at age r, provided that the member survives to age r. Further, the retirement benefit is 1 unit initially, with annual increase as governed by the function h. The denominator is the actuarial present value of the payment at unit rate per annum for the entire service period starting from entry age a to retirement age r. Thus $P(x)$ is analogous to the net benefit premium. In this case the projected benefit is funded by a level contribution from entry age to retirement, and hence this accrued cost method is popularly known as "entry age normal cost method", cf. (Anderson, 1985).

In the entire discussion above, age x is treated as a continuous variable. In the numerical illustration we will treat x as a discrete variable, and hence in the following, we define appropriately $M(x), m(x)$, and $P(x)$ in the discrete setup. Taking clue from $M(x) = \frac{\bar{a}_{a:\overline{x-a}|}}{\bar{a}_{a:\overline{r-a}|}}$, $a < x < r$, in discrete setup, we define $M(x) = \frac{\ddot{a}_{a:\overline{x-a}|}}{\ddot{a}_{a:\overline{r-a}|}}$, $x = a, a+1, \ldots, r$. It is to be noted with this definition of $M(x)$, we get $M(a) = 0$, as it should be, and $M(r) = 1$. We simplify $M(x)$ for $x = a+1, \ldots, r$ as follows:

$$M(x) = \frac{\ddot{a}_{a:\overline{x-a}|}}{\ddot{a}_{a:\overline{r-a}|}} = \frac{\sum_{k=1}^{x-a} v^k {}_k p_a}{\sum_{k=1}^{r-a} v^k {}_k p_a}$$

$$= \frac{\sum_{k=1}^{x-a} v^k S(a+k)}{\sum_{k=1}^{r-a} v^k S(a+k)} = \frac{\sum_{y=a+1}^{x} v^{y-a} S(y)}{\sum_{y=a+1}^{r} v^{y-a} S(y)}$$

$$= \frac{\sum_{y=a+1}^{x} v^y S(y)}{\sum_{y=a+1}^{r} v^y S(y)}, \quad x = a+1, \ldots, r.$$

The function $m(x)$, which we will interpret as the probability mass function corresponding to $M(x)$ defined above, is given by

$$m_1(x) = \frac{v^x S(x)}{\sum_{y=a+1}^{r} v^y S(y)}, \quad x = a+1, \ldots, r.$$

In a discrete setup, $P(x)$ corresponding to this $m_1(x)$ is defined as

$$P(x) = (aA)(x)m_1(x) = v^{r-x} \frac{S(r)}{S(x)} \ddot{a}_r^{(h)} \frac{v^x S(x)}{\sum_{y=a+1}^{r} v^y S(y)} = \frac{v^r S(r) \ddot{a}_r^h}{\sum_{y=a+1}^{r} v^y S(y)}.$$

It is to be noted that $P(x)$ is again free from x. the following example evaluates $P(x)$ for $m_1(x) = \frac{v^x S(x)}{\sum_{y=a+1}^{r} v^y S(y)}, x = a+1, \ldots, r$, for the specific benefit function.

Example 4.2.3 Suppose that the retirement age is $r = 65$, entry age is 25, and the survival function is given by $S(x) = \exp\{-m(c^{x-25} - 1)\}, x \geq 25$, where $m = B/\log_e C$. Assume that $B = 0.001$ and $C = 1.098$. Suppose that $h(x) = e^{\eta(x-r)}$, $x \geq r$, $\eta = 0.045$, $\delta = 0.05$.

(i) Find $\ddot{a}_{65}^{\eta}$.
(ii) Find $P(x)$ in the discrete setup when the accrual function is $m_1(x) = \frac{v^x S(x)}{\sum_{y=a+1}^{r} v^y S(y)}, x = a+1, \ldots, r$.
(iii) Find the normal cost per annum if the initial pension benefit is Rs 300000/-. Also, find the corresponding $(aV)(x)$.

Solution In the discrete setup, $P(x) = 300000 \frac{e^{-\delta r} S(r) \ddot{a}_r^{\eta}}{\sum_{y=a+1}^{r} S(y) v^y}$, which is free from x for the given accrual function. Further,

$$(aV)(x) = (aA)(x)M(x) = P(x)M(x)/m(x).$$

The following is a set of R commands to compute $\ddot{a}_{65}^{\eta}$, $P(x)$, and $(aV)(x)$:

```
a <- 1.098   # C;
b <- 0.001   # B;
m <- b/log(a, base=exp(1));
e <- exp(1);
del <- 0.05;
eta <- 0.045;
x <- 65:100;
p <- e^(m-m*a^(x-25))   #survival function S(x);
v <- e^(-(del-eta));
ad <- v^(-65)*sum(v^x*p)/p[1];
ad;
x <- 26:65;
p <- e^(m-m*a^(x-25))   #survival function S(x);
v <- e^(-del);
```

Table 4.1 Accrued actuarial liability

Age $(x, x+1)$	$(aV)(x)$	Age $(x, x+1)$	$(aV)(x)$
25–26	17427.41	45–46	661650.76
26–27	35769.43	46–47	718214.56
27–28	55078.33	47–48	778680.90
28–29	75410.06	48–49	843434.04
29–30	96824.70	49–50	912912.26
30–31	119386.84	50–51	987617.45
31–32	143166.10	51–52	1068126.63
32–33	168237.64	52–53	1155105.87
33–34	194682.88	53–54	1249327.34
34–35	222590.14	54–55	1351689.96
35–36	252055.57	55–56	1463244.90
36–37	283184.06	56–57	1585226.93
37–38	316090.40	57–58	1719093.37
38–39	350900.58	58–59	1866572.61
39–40	387753.32	59–60	2029724.99
40–41	426801.83	60–61	2211019.69
41–42	468215.88	61–62	2413432.37
42–43	512184.21	62–63	2640570.08
43–44	558917.43	63–64	2896832.30
44–45	608651.30	64–65	3187620.00

```
y1 <- v^(65)*p[40]*ad;
y2 <- sum(p*v^x);
y3 <- y1/y2;
y4 <- 300000*y3;
y3;   y4;
p1 <- p*v^x/y2   # m(x);
p2 <- cumsum(p1)   # M(x);
p3 <- y4*p2/p1   # (aV)(x);
d <- data.frame(x, p3);
d   #Table 4.1;
```

For the given mortality and interest pattern, $\ddot{a}_{65}^{\eta} = 10.6254$. For the given data, if the initial pension benefit is 1 unit, the normal cost is 0.058091 and if the initial annual pension benefit is Rs 300000/-, then the normal cost is 17427.41 for each unit age group $(x, x+1)$ credited to the pension fund at the end of the year. In other words, if Rs 17427.41 is credited to the pension fund annually after completing one year in the plan, that is, from age 26 to 65, the individual will receive initial annual pension of Rs 300000/-, and for future years, the pension will increase at the rate $e^{\eta} = 1.046$ per year. Table 4.1 gives the values of accrued actuarial liability for all ages. $(aV)(r) = 3187620.00$ represents the accrued liability at the retirement age r, that is, it represents the fund that will be sufficient to provide pension payments with initial benefit of 300000 with annual increase at the rate of e^{η}. Thus, $(aV)(r)$ should equal $300000\ddot{a}^{\eta}_{65}$. It is to be noted that

$$300000\ddot{a}^{\eta}_{65} = 300000 \times 10.6254 = 3187620.00.$$

In Examples 4.2.2 and 4.2.3, we have noted that the normal cost rate is the same for all ages when the accrual function is $m_1(x) = \dfrac{S(x)e^{-\delta x}}{\int_a^r S(y)e^{-\delta y}\,dy}$ or $m_1(x) = \dfrac{S(x)v^x}{\sum_{a+1}^r S(y)v^y}$. With this accrual function the retirement benefit is not related to salary scale, and consequently contribution to the pension fund is also not related to salary. The salary of members changes due to individual experience and merit and also due to inflation and changes in the productivity of all the participants. To incorporate these factors of changes in salary, as in Chap. 3, let

$$w_x = (1+0.05)^{x-a}u_x, \quad a \le x \le r,$$

denote the salary scale for each member of age x. u_x indicates the salary change due to individual experience and merit, and $(1+0.05)^{x-a}$ indicates change in salary scale by a year-of-experience factor. Suppose that $e^{\tau t}$ reflects inflation and changes in the productivity of all the participants at time t. This factor does not depend on the age of the individual. Thus the annual salary rate expected at time t by a member of age x is given by $w_x e^{\tau t}$, $a \le x \le r$. In Chap. 3 we have taken $\tau = 0$.

If the initial retirement benefit is related to the salary scale, then it seems reasonable to have the contribution to pension fund as a fraction of salary rate at that age. To incorporate such feature involving rate of salary, we define the accrual function $m_2(x)$ proportional to $S(x)e^{-\delta x}e^{\tau t}w_x$. We normalize this function so that $\int_a^r m_2(y)\,dy = 1$. If the member is of age x at time t, then the member's age was $y < x$ at time $t - (x-y) = t+y-x$ and will be $y > x$ at time $t+(y-x) = t+y-x$. Thus, for $a < y < r$, $m_2(y)$ is proportional to $S(y)e^{-\delta y}e^{\tau(t+y-x)}w_y$. Thus, the normalized $m_2(x)$ is given by

$$m_2(x) = \frac{S(x)e^{-\delta x}e^{\tau t}w_x}{\int_a^r S(y)e^{-\delta y}e^{\tau(t+y-x)}w_y\,dy} = \frac{S(x)e^{-\delta x}e^{\tau x}w_x}{\int_a^r S(y)e^{-\delta y}e^{\tau y}w_y\,dy}.$$

The following example derives the formula of $P(x)$ and shows how the salary scale affects the contribution rate for each age.

Example 4.2.4 Let the accrual function $m_2(x)$ be defined as

$$m_2(x) = \frac{S(x)e^{-\delta x}w_x e^{\tau x}}{\int_a^r S(y)e^{-\delta y}w_y e^{\tau y}\,dy}.$$

Find the normal constant rate $P(x)$.

Solution By definition,

$$P(x) = (aA)(x)m(x)$$

$$= e^{-\delta(r-x)}\frac{S(r)}{S(x)}\bar{a}_r^h \times \frac{S(x)e^{-\delta x}e^{\tau x}w_x}{\int_a^r S(y)e^{-\delta y}e^{\tau y}w_y\,dy}$$

$$= \frac{e^{-\delta r}S(r)\bar{a}_r^h e^{\tau x}w_x}{\int_a^r S(y)e^{-\delta y}e^{\tau y}w_y\,dy}$$

$$= kw_x e^{\tau x},$$

where k is free from x but depends on the force of interest, the actuarial present value of the annuity based on the function h, and the value of survival function. In this case the contribution rate is proportional to the salary scale function. This method is also known as entry age normal cost method with contribution proportional to the salary scale, cf. (Anderson, 1985).

In a discrete setup, $m_2(x)$ is defined analogously as

$$m_2(x) = \frac{v^x S(x)w_{x-1}e^{\tau(x-1)}}{\sum_{y=a+1}^r v^y S(y)w_{y-1}e^{\tau(y-1)}}, \quad x = a+1,\ldots,r.$$

The normal cost $P(x)$ corresponding to $m_2(x)$ is given by

$$P(x) = \frac{v^r S(r)\ddot{a}_r^\eta w_{x-1}e^{\tau(x-1)}}{\sum_{a+1}^r S(y)v^y w_{y-1}e^{\tau(y-1)}}.$$

The following example computes $P(x)$ and $(aV)(x)$ for $m_2(x) = \frac{v^x S(x)w_{x-1}e^{\tau(x-1)}}{\sum_{y=a+1}^r v^y S(y)w_{y-1}e^{\tau(y-1)}}, x = a+1,\ldots,r.$

Example 4.2.5 Suppose that the retirement age is $r = 65$, the entry age is 25, and the survival function is given by $S(x) = \exp\{-m(c^{x-25}-1)\}$, $x \geq 25$, where, $m = B/\log_e C$. Assume that $B = 0.001$ and $C = 1.098$. Suppose that $h(x) = e^{\eta(x-r)}$, $x \geq r$, $\eta = 0.045$, $\delta = 0.05$. Suppose that the salary rate function w_x is as given in column 7 of Table 3.3 in Chap. 3 and $\tau = 0.02$. Further, it is given that $\ddot{a}_{65}^\eta = 10.6254$. In the discrete setup, find the constant k, as defined in Example 4.2.4, the normal cast $P(x)$ and $(aV)(x)$ for $m_2(x) = \frac{v^x S(x)w_{x-1}e^{\tau(x-1)}}{\sum_{y=a+1}^r v^y S(y)w_{y-1}e^{\tau(y-1)}}, x = a+1,\ldots,r$, if the initial annual pension benefit is Rs 300000/-, with annual increase of e^η.

Solution In the discrete setup, $P(x) = \frac{v^r S(r)\ddot{a}_r^\eta w_{x-1}e^{\tau(x-1)}}{\sum_{a+1}^r S(y)v^y w_{y-1}e^{\tau(y-1)}}$. The following is a set of R commands to compute $P(x)$ and $(aV)(x)$, and the values are displayed in Table 4.2.

```
a <- 1.098    # C;
b <- 0.001    # B;
```

Table 4.2 Age wise normal cost rate and accrued actuarial liability

Age $(x, x+1)$	$P(x)$	$(aV)(x)$	Age $(x, x+1)$	$P(x)$	$(aV)(x)$
25–26	5363.70	5363.70	45–46	22564.78	399806.88
26–27	5745.65	11390.85	46–47	24245.12	447700.43
27–28	6196.68	18186.69	47–48	26067.43	500596.26
28–29	6663.58	25809.27	48–49	28038.43	559059.32
29–30	7146.82	34320.73	49–50	29991.74	623552.27
30–31	7706.14	43846.95	50–51	32277.83	694952.91
31–32	8224.67	54404.44	51–52	34644.04	773985.45
32–33	8884.40	66193.69	52–53	37092.69	861476.77
33–34	9504.48	79246.26	53–54	39813.93	958563.62
34–35	10274.42	93786.54	54–55	42629.97	1066361.31
35–36	11006.07	109864.73	55–56	45641.28	1186260.71
36–37	11829.93	127666.62	56–57	48756.85	1319780.57
37–38	12750.77	147395.70	57–58	52081.40	1468803.28
38–39	13634.43	169135.86	58–59	55520.37	1635440.68
39–40	14690.52	193189.49	59–60	59077.03	1822198.02
40–41	15783.71	219745.43	60–61	62646.72	2031957.61
41–42	16915.08	249010.39	61–62	66446.72	2268408.35
42–43	18236.43	281362.03	62–63	70487.11	2536004.19
43–44	19604.26	317065.07	63–64	74778.31	2840159.52
44–45	21019.91	356411.94	64–65	79448.13	3187620.00

```
m <- b/log(a, base=exp(1));
ad <- 10.6254;
e <- exp(1);
x <- 26:65;
p <- e^(m-m*a^(x-25))   #survival function S(x);
del <- 0.05;
v <- e^(-del);
tau <- 0.02;
v1 <- e^(tau);
z <- read.table("D:service.txt", header=T);
z1 <- z[, 2]   #values of service function w_x;
y1 <- v^(65)*p[40]*ad*v1^(x-1)*z1;
y2 <- sum(p*v^x*v1^(x-1)*z1);
y3 <- y1/y2;
y4 <- 300000*y3;
k <- v^(65)*p[40]*ad/y2;
k;
p1 <- p*v^x*v1^(x-1)*z1/y2   # m(x);
p2 <- cumsum(p1)   # M(x);
```

```
p3 <- y4*p2/p1   # (aV)(x);
d <- data.frame(x, y4, p3);
d   # Table 4.2;
```

For the given setup, $k = 0.010844$. The proportion k of annual salary rate $w_x e^{\tau x}$ is the contribution to the pension fund. In Example 4.2.3 the normal cost is constant given by 17427.41 for all the ages. In this example we see that the contribution rate increases as age increases due to the increase in salary scale. Thus, if amount $P(x)$ is credited to the pension fund from age 25 to 65, the fund will accumulate to provide the individual retirement benefit with initial annual pension of Rs 300000/- and annual increase at the rate e^{η} per year. The third and sixth columns, specifying the values of $(aV)(x)$, present the accumulation of the fund. It is accumulated to 3187620.00 at age 65, which is exactly equal to $300000\ddot{a}_{65}^{\eta}$.

Examples 4.2.3 and 4.2.5 depict the role of the accrual function, to decide the annual contribution to the pension fund sufficient to provide the retirement benefits and the accrued actuarial liability. The rate of contribution and the accrued actuarial liability change as the accrual function changes. In the following we discuss the accrued actuarial liability $(aV)(x)$ in some more details. We have defined the accrued actuarial liability as

$$(aV)(x) = (aA)(x)M(x).$$

The concept of accrued actuarial liability is more or less similar to the concept of reserve in life insurance. The above definition of $(aV)(x)$ is analogous to the retrospective reserve. With the prospective approach, it can be defined as the actuarial present value of the pension benefit at age x less the actuarial present value of future normal costs. With this approach it is defined as

$$(aV)(x) = (aA)(x) - (Pa)(x), \quad \text{where } (Pa)(x) = \int_x^r e^{-\delta(y-x)} \frac{S(y)}{S(x)} P(y) \, dy.$$

$(Pa)(x)$ denotes the actuarial present value of future normal costs. Thus, $(Pa)(x)$ can be expressed as

$$(Pa)(x) = (aA)(x) - (aV)(x) = (aA)(x)\big[1 - M(x)\big].$$

The accrued actuarial liability is thus defined as

$$(aV)(x) = (aA)(x)M(x) = (aA)(x) - (Pa)(x).$$

The following example verifies that for accrual function $m_2(x)$, these two approaches, prospective and retrospective, of $(aV)(x)$ result in the same value.

Example 4.2.6 Prove that $(aV)(x) = (aA)(x)M(x) = (aA)(x) - (Pa)(x)$ if the accrual function is given by

$$m_2(x) = \frac{S(x)e^{-\delta x}e^{\tau x}w_x}{\int_a^r S(y)e^{-\delta y}e^{\tau y}w_y \, dy}.$$

Solution We have $P(x) = \dfrac{e^{-\delta r} S(r)\bar{a}_r^h w_x e^{\tau x}}{\int_a^r S(y)e^{-\delta y}e^{\tau y}w_y \, dy}$, which we write as $P(x) = \dfrac{e^{-\delta r} S(r)\bar{a}_r^h w_x e^{\tau x}}{I}$, where $I = \int_a^r S(y)e^{-\delta y}e^{\tau y}w_y \, dy$. By prospective approach of the definition of reserve,

$$(aV)(x) = (aA)(x) - (Pa)(x)$$

$$= (aA)(x) - \int_x^r e^{-\delta(y-x)} \frac{S(y)}{S(x)} P(y) \, dy$$

$$= e^{-\delta(r-x)} \frac{S(r)}{S(x)} \bar{a}_r^h - (1/I) \int_x^r e^{-\delta(y-x)} \frac{S(y)}{S(x)} e^{-\delta r} S(r)\bar{a}_r^h w_y e^{\tau y} \, dy$$

$$= e^{-\delta(r-x)} \frac{S(r)}{S(x)} \bar{a}_r^h \left\{ 1 - \left[\frac{\int_x^r e^{-\delta y} S(y)e^{\tau y} w_y \, dy}{\int_a^r e^{-\delta y} S(y)w_y e^{\tau y} \, dy} \right] \right\}$$

$$= e^{-\delta(r-x)} \frac{S(r)}{S(x)} \bar{a}_r^h \left\{ \frac{\int_a^x e^{-\delta y} S(y)w_y e^{\tau y} \, dy}{\int_a^r e^{-\delta y} S(y)w_y e^{\tau y} \, dy} \right\}$$

$$= (aA)(x) \int_a^x m(y) \, dy = (aA)(x)M(x),$$

which is the formula for retrospective reserve.

We have discussed how the normal cost rate for the individual is to be obtained using two accrual functions. A similar procedure is adopted for various forms of accrual function $m(x)$. Employer has to set aside the normal cost for all the members of all ages between a to r in the plan. The total normal cost rate, say $NC(t)$, that employer has to pay in the pension fund at time t is obtained by summing it over all the members $n(x, t)$ of age x in the group at time t.

We now proceed to discuss the accrued benefit cost method for a group to obtain the total normal cost rate at a specific time point for a group consisting of members of all ages.

4.3 Accrued Benefit Cost Method for a Group

The basic component in this approach is recognizing the pension obligations for a group of members retiring at time t. Toward it we first discuss a terminal funding method and then its application to the accrued benefit cost method for a group.

Terminal Funding Method   Under the terminal funding approach, the employer sets aside for each employee, on the date of retirement of an employee, a lump-sum amount sufficient to provide the monthly pension benefit promised under the plan. This amount is essentially a single premium or the purchasing price of the monthly whole life annuity due from an insurance company. Thus, the liability is discharged in full when the employee retires, and no assets are set aside while the employee is working. Under this method, the required contribution rate, referred to as normal cost rate, is denoted by $^{T}P_t$. The presuperscript stands for T in the nomenclature "Terminal Funding method." It is a rate at which the actuarial present value of future pensions for members reaching retirement age r is incurred at time t. Thus, in this approach the cost of the plan is recognized only at the time of retirement of the members. Other extreme of terminal funding method is the initial funding method in which the pension liability is recognized when the person joins the employer. Both these methods are not used in practice. A method which recognizes the pension liability periodically, in particular, annually, during the active service period of the member is used in practice. In all these methods, $^{T}P_t$ is a basic building block for the various functions used to describe the funding operations. Hence, we proceed to find the expression of $^{T}P_t$. To find the expression for $^{T}P_t$, we define the functions to take into account the changing size of the working group and the changes in salary due to a variety of factors, explained in Chap. 3. To emphasize the funding methods, we restrict to the age-service retirement at age r.

We consider a group consisting of members entering at age a, retiring at age r, and subject to the survival function $S(x)$ with $S(a) = 1$. For $a < x < r$, we further assume that decrements occur due to mortality or other causes as specified in Chap. 3, but for $x > r$, mortality is the only cause of decrement. Thus, in our usual notation, $\frac{S(x)}{S(a)} = {}_{x-a}p_a^{(\tau)}$, $x \le r$, is the probability that the individual joining the group of employees is still in the group at age x. Let $n(a, u)$ denote the number of new entrants at age a at time u. The number of members attaining age x at time t is then given by

$$n(x, t) = n(a, t - x + a) {}_{x-a}p_a^{(\tau)} = n(a, t - x + a)\frac{S(x)}{S(a)} = n(a, t - x + a)S(x),$$

as $S(a) = 1$. $n(a, t - x + a)$ denotes the number of participants of age a at time $u = t - (x - a)$. Thus, $n(a+1, t) = n(a, t-1)S(a+1)$, the number of individuals of age $a + 1$ at time t, will be the number of individuals of age a at time $t - 1$ who survive for a year. Similarly $n(a+2, t) = n(a, t-2)S(a+2)$, and so on. We further assume that the population is closed and there is no in migration, in the sense that there are no new entrants in the group beyond age a and population of working group decreases due to death, disability, withdrawal or death. This assumption remains valid in practice if we consider a group of individuals joining at later ages as a different group, separately for each age. The progress of such groups can be studied on similar lines.

If $n(x, t)$ is independent of t, we then say that the population is stationary. In this case we denote $n(x, t)$ by l_x. In Chap. 3, in the service table, Table 3.3, we have assumed that the population is stationary. In single and multiple decrement tables discussed in Chaps. 1 and 2, we have assumed that the population is stationary, although it is rarely true. In stationary population it is assumed that exactly l_0 births occur in each calender year or the number of new entrants in a group is l_a for each year. The number of decrements at various ages due to various causes also remain the same over the years. Further, in such a population, age composition never changes, and the total size remains constant. In reality, the number of births or the number of entrants changes, and the decrement pattern changes from year to year, and hence the populations are rarely stationary. A restricted amount of variation from a stationary population is introduced by assuming that the number of births or new entrants increases at a constant rate R, which can be positive or negative. It is still assumed that the age-specific decrement rates remain the same. As a consequence, the total size of the population increases at a rate e^R per year, but the composition of the population by age remains constant. Such a population is known as a stable population. Thus, if $n(x, t)$ is of the form $e^{Rt} l_x$, where R is a constant, the population is known as a stable population. If $R = 0$, then a stable population is a stationary population. If $R > 0$, then the population grows exponentially, and if $R < 0$, then the population decreases exponentially. The case $R > 0$ reflects the period of boom when the population in working period $(25, 65)$ is higher than that in the stationary population for each age, while the case $R < 0$ reflects the period of depression when the population in working period $(25, 65)$ is less than that in the stationary setup. For a stable population, the fraction of the total population that lies between ages x_0 and x_1 is independent of t. Thus, the size of a stable population changes over time, but its relative age distribution is constant, that is why it is labeled as the stable population. The conditions for a stable or stationary population are seldom realized in practice in view of the changes in survival function and density of births or new entrants in a group under study. We use these models to illustrate the theoretical developments.

The salary of members changes due to individual experience and merit and also due to inflation and changes in the productivity of all the participants. The annual salary rate expected at time t by a member of age x is given by $w_x e^{\tau t}$, $a \le x \le r$. With these functions the total salary rate at time t for $n(x, t)$ members of age x is given by

$$W(x, t) = n(x, t) w_x e^{\tau t}.$$

In particular, $W(r, t) = n(r, t) w_r e^{\tau t}$ denotes the total salary rate at the time of retirement of the individual of age r at time t. The initial pension benefit is, as discussed in Chap. 3, a fraction of $W(r, t)$. Thus the function $W(r, t)$ is an important component in the expression of ${}^T P_t$.

At any time the group of employees consists of individuals of all ages from a to r. Hence the total annual salary rate W_t at time t is given by

$$W_t = e^{\tau t} \int_a^r n(x, t) w_x \, dx.$$

If age x is taken as a discrete variable, then

$$W(x, t) = n(x, t) w_x e^{\tau t}, \quad x - a, \ldots, r - 1.$$

W_t is obtained by summing over all ages between x and $r - 1$. Thus in discrete setup, W_t is given by,

$$W_t = \sum_{x=a}^{r-1} W(x, t) = \sum_{x=a}^{r-1} n(x, t) w_x e^{\tau t} = e^{\tau t} \sum_{a}^{r-1} n(x, t) w_x.$$

The upper limit in the sum is $r - 1$ in view of the fact that w_x presents the salary scale for an individual of complete age x, which remains the same for one year. If the retirement age is taken as 65, then the salary scale for the last year is w_{64}.

W_t represents the total payroll payment rate at time t. It conveys the employer the expected amount of total salary rate at any time. Contribution to the pension fund on behalf of all employees at time t, is a fraction of this total payroll payment rate. The following example illustrates the changes in population sizes in stationary and stable setup and the computation of total payment rate.

Example 4.3.1 Suppose that the total force of decrement is given by $\mu_x = BC^{x-25}$, $x \geq 25$. Assume that $B = 0.001$, $C = 1.098$, $l_0 = 1000$, and the salary scale function $w_x = (1.05)^{x-25} u_x$ as given in column 7 of Table 3.3 in Chap. 3. Let $\tau = 0.02$.

(i) Find l_x for a stationary population and $n(x, t)$ for a stable population with $R = 0.02$ and $R = -0.01$, and for $t = 10, 20$, and 30. Plot the graphs of population sizes.
(ii) Tabulate the population sizes for ages 25 to 65.
(iii) Find W_t for $t = 10, 20$, and 30, for the stationary and the stable population when the retirement age is 65.

Solution For a stable population, $n(x, t) = e^{Rt} l_x$, where $l_x = l_0 S(x)$. For the given force of mortality, the survival function is given by $S(x) = \exp\{-m(c^{x-25} - 1)\}$, $x \geq 25$, where $m = B / \log_e C$. Hence,

$$W_t = e^{0.02t} \sum_{25}^{64} n(x, t) w_x = e^{0.02t} \sum_{25}^{64} e^{Rt} l_x w_x = e^{(0.02+R)t} \sum_{25}^{64} l_x w_x.$$

The following is a set of R commands to compute the population sizes, their graphs, and total payroll payment rate;

```
a <- 1.098    # C;
b <- 0.001    # B;
m <- b/log(a, base=exp(1));
e <- exp(1);
x <- 25:100;
p <- e^(m-m*a^(x-25))    # survival function S(x);
y <- 1000*p  # l_x;
r <- 0.02;
t <- c(10, 20, 30);
ys1 <- e^(r*t[1])*y;
```

```
ys2 <- e^(r*t[2])*y;
ys3 <- e^(r*t[3])*y;
par(mfrow=c(1, 1), font.axis=2, font.lab=2, cex.axis=1,
    cex.lab=1.5, font=2, lwd=2);
plot(x, y, "o", pch=20, cex=0.7, main=" ", xlab=" ",
    ylab=" ", ylim=range(0, max(ys3)));
lines(x, ys1, "o", pch=15, cex=0.7, main=" ");
lines(x, ys2, "o", pch=24, cex=0.7, main=" ");
lines(x, ys3, "o", pch=3, cex=0.7, main=" ");
legend(locator(1), pch=c(20, 15, 24, 3), legend=c("R=0",
        "t=10, R=0.02", "t=20, R=0.02", "t=30, R=0.02"),
        cex=1.2);
ys <- round(data.frame(y, ys1, ys2, ys3), 2);
d <- data.frame(x, ys);
d1 <- d[1:41, ];
d1   # Table 4.3;
```

In the above set of commands replacing $r = 0.02$ by $r = -0.01$, we get Fig. 4.2 and Table 4.4. The following commands compute W_t:

```
z <- read.table("D:service.txt", header=T);
z1 <- z[, 2]   # values of w_x;
y1 <- y[1:40]   # l_x for x = 25 to 64;
y2 <- sum(y1*z1);
tau <- 0.02   # value of τ;
r <- c(0.02, 0, -0.01);
w1 <- e^((tau+r)*t[1])*y2;
w2 <- e^((tau+r)*t[2])*y2;
w3 <- e^((tau+r)*t[3])*y2;
w <- data.frame(r, w1, w2, w3);
w   # Table 4.5;
```

From Table 4.3 we note that the number of working individuals at any age in stable population is higher than the corresponding number in stationary population for all the three time points. Within the stable population, as time t increases, the population size also increases as $R > 0$. Similar pattern is revealed in Fig. 4.1. In Table 4.4, we see the reverse scenario. The number of working individuals at any age in stable population is less than the corresponding number in stationary population for all the three time points and the stable population decreases as time t increases, as $R < 0$. Figure 4.2 depicts the same scenario. For the given force of mortality and the given set of parameters, the population size approaches 0 in all the cases after age 90. So it is a reasonably good model for the population size.

Table 4.5 reports the total payment rate for three time epochs and for three types of populations.

As expected, W_t increases as t increases for all the three types of populations, due to factor $e^{\tau t}$. For a stable population with $R = -0.01$, W_t is the smallest for

Table 4.3 Population size
with $R = 0$ and $R = 0.02$

Age x	Stationary	$t = 10$	$t = 20$	$t = 30$
25	1000.00	1221.40	1491.82	1822.12
26	998.95	1220.12	1490.26	1820.21
27	997.80	1218.72	1488.55	1818.12
28	996.54	1217.18	1486.67	1815.82
29	995.16	1215.49	1484.61	1813.30
30	993.65	1213.64	1482.35	1810.54
31	991.99	1211.61	1479.87	1807.51
32	990.16	1209.39	1477.15	1804.20
33	988.17	1206.95	1474.18	1800.56
34	985.98	1204.28	1470.92	1796.58
35	983.59	1201.36	1467.34	1792.22
36	980.97	1198.16	1463.43	1787.44
37	978.10	1194.65	1459.15	1782.21
38	974.95	1190.81	1454.46	1776.48
39	971.51	1186.61	1449.33	1770.21
40	967.75	1182.01	1443.71	1763.36
41	963.64	1176.99	1437.57	1755.86
42	959.14	1171.49	1430.87	1747.66
43	954.22	1165.49	1423.53	1738.71
44	948.86	1158.94	1415.53	1728.93
45	943.00	1151.78	1406.79	1718.25
46	936.61	1143.97	1397.25	1706.61
47	929.64	1135.47	1386.86	1693.92
48	922.05	1126.19	1375.54	1680.09
49	913.79	1116.10	1363.21	1665.03
50	904.80	1105.13	1349.80	1648.65
51	895.03	1093.20	1335.24	1630.86
52	884.43	1080.25	1319.42	1611.54
53	872.94	1066.21	1302.27	1590.60
54	860.49	1051.00	1283.69	1567.91
55	847.02	1034.55	1263.61	1543.37
56	832.48	1016.79	1241.91	1516.87
57	816.80	997.64	1218.52	1488.30
58	799.92	977.02	1193.34	1457.54
59	781.79	954.88	1166.29	1424.51
60	762.35	931.14	1137.30	1389.10
61	741.57	905.75	1106.29	1351.23
62	719.40	878.68	1073.22	1310.83
63	695.82	849.88	1038.05	1267.87
64	670.82	819.35	1000.75	1222.32
65	644.41	787.08	961.34	1174.19

Table 4.4 Population size with $R = 0$ and $R = -0.01$

Age x	Stationary	$t = 10$	$t = 20$	$t = 30$
25	1000.00	904.84	818.73	740.82
26	998.95	903.89	817.87	740.04
27	997.80	902.85	816.93	739.19
28	996.54	901.71	815.90	738.26
29	995.16	900.46	814.77	737.23
30	993.65	899.09	813.53	736.11
31	991.99	897.59	812.17	734.88
32	990.16	895.94	810.68	733.53
33	988.17	894.13	809.04	732.05
34	985.98	892.16	807.26	730.43
35	983.59	889.99	805.29	728.66
36	980.97	887.62	803.15	726.72
37	978.10	885.02	800.80	724.59
38	974.95	882.17	798.22	722.26
39	971.51	879.06	795.41	719.71
40	967.75	875.66	792.33	716.93
41	963.64	871.93	788.96	713.88
42	959.14	867.86	785.28	710.55
43	954.22	863.42	781.25	706.91
44	948.86	858.56	776.86	702.93
45	943.00	853.26	772.06	698.59
46	936.61	847.48	766.83	693.86
47	929.64	841.17	761.13	688.69
48	922.05	834.31	754.91	683.07
49	913.79	826.83	748.15	676.95
50	904.80	818.70	740.79	670.29
51	895.03	809.86	732.79	663.06
52	884.43	800.27	724.11	655.20
53	872.94	789.87	714.70	646.69
54	860.49	778.60	704.51	637.46
55	847.02	766.42	693.48	627.49
56	832.48	753.26	681.57	616.71
57	816.80	739.07	668.74	605.10
58	799.92	723.80	654.92	592.59
59	781.79	707.39	640.07	579.16
60	762.35	689.80	624.16	564.76
61	741.57	671.00	607.15	549.37
62	719.40	650.94	589.00	532.95
63	695.82	629.61	569.69	515.48
64	670.82	606.99	549.22	496.96
65	644.41	583.09	527.60	477.39

Fig. 4.1 Stationary and
stable population with
$R = 0.02$

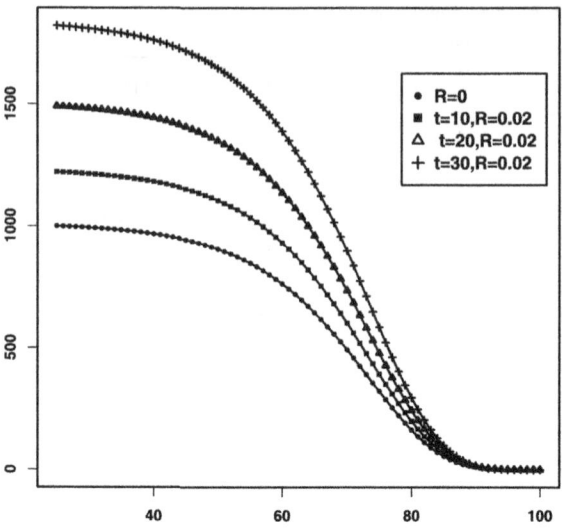

Fig. 4.2 Stationary and
stable population with
$R = -0.01$

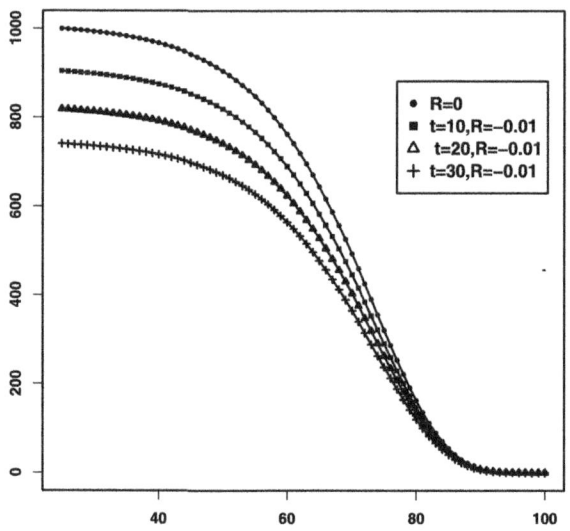

all t, while for a stable population with $R = 0.02$, W_t is the highest for all t. The
fact that for $R = 0$ with $t = 10$, W_t is same as that for $R = -0.01$ with $t = 20$,
follows easily from the formula of W_t.

To obtain the expression for ${}^T P_t$ when age x is taken as a discrete variable, we
begin with a pension plan which provides retirement annuities payable only after
retirement age r and the initial annual pension rate is a fraction d of the salary rate
at retirement. Thus, for a member retiring at time t, the projected initial pension
payment rate is $d w_{r-1} e^{\tau t}$, where d is a fraction, and $w_{r-1} e^{\tau t}$ is a salary rate at retirement. The pension payment does not remain the same for all the future years but

Table 4.5 Total payroll
payment rate

R	$t = 10$	$t = 20$	$t = 30$
0.02	162864.90	242965.90	362462.50
0	133342.50	162864.90	198923.60
−0.01	120653.30	133342.50	147366.30

increases periodically, usually annually. To incorporate this increase, as in Chap. 3, suppose that $h(x)$, with $h(r) = 1$, denotes an adjustment factor applied to the initial pension payment rate $dw_{r-1}e^{\tau(t-x+r)}$ for those who retired $x - r$ years ago. For a retiree of age x, $x \geq r$, at time t, the annual rate of pension payment is projected as $dw_{r-1}e^{\tau(t-x+r)}h(x)$. To obtain the expression for ^TP_t, we need to find the actuarial present value of the future pension benefits starting at age r. Let δ denote the force of interest, and $\bar{a}_r^h$ denote the actuarial present value of the life annuity at age r of the pension benefit which takes into account the increase in pension as governed by the function $h(x)$. Suppose $h(x) = \exp[\eta(x - r)]$, where η is a constant rate of increase, possibly related to the expected inflation rate. We have proved that

$$\bar{a}_r^\eta = \int_r^\infty e^{-\delta(x-r)}h(x)\frac{S(x)}{S(r)}\,dx = \bar{a}_r\left(\delta'\right),$$

where $\delta' = \delta - \eta$, that is, $\bar{a}_r^\eta$ is the annuity function with force of interest $\delta' = \delta - \eta$. Further, $n(r, t)$ members attaining age r at time t will collect pension at an average initial rate of $dw_{r-1}e^{\tau t}$. With the annual increase at rate e^η, the expression for ^TP_t is given by

$$^TP_t = dw_{r-1}e^{\tau t}n(r, t)\bar{a}_r^\eta.$$

In Chap. 3, we have defined a function $R(a, h, t)$ as the initial benefit if the individual joins the group at age a, is in active service at age $a + h$, and retires at age $a + h + t = r$ in this setup. If we denote such an initial benefit at age r by $b(r)$, then the terminal fund at time t is defined as

$$^TP_t = b(r)e^{\tau t}n(r, t)\bar{a}_r^\eta.$$

For a stable population, $n(r, t) = l_r e^{Rt}$. Hence, for a stable population, ^TP_t is given by

$$^TP_t = dw_{r-1}\bar{a}_r^\eta l_r e^{(\tau+R)t}.$$

Remark 4.3.1 The annuity function $\bar{a}_r^\eta$ in the above expression may be replaced by $\ddot{a}_r^\eta$ or $\ddot{a}_r^{(12)\eta}$, as most of the pension payments are monthly, that is, according to discrete annuity due. $\ddot{a}_r^\eta$ is given by

$$\ddot{a}_r^\eta = \sum_{x=r} v^{x-r}\,_{x-r}p_r = \sum_{x=r} v^{x-r}\frac{S(x)}{S(r)}, \quad \text{where } v = e^{-\delta'}.$$

Table 4.6 Normal cost rate in terminal funding

t	$R = 0.02$	$R = 0$	$R = -0.01$
10	8322.94	6814.24	6165.78
20	12416.36	8322.94	6814.24
30	18523.03	10165.66	7530.90

Continuous annuities are mathematically simple to use, and these are usually good approximation to the monthly annuity due; hence continuous annuities are frequently used in theoretical development.

In the following example we compute $^T P_t$ for the setup of Example 4.3.1 in two cases. First, we assume that the initial annual pension benefit is dw_{r-1} and then take the initial annual benefit as obtained in Example 3.2.5 in Chap. 3. In Example 3.2.5, we have obtained the projected annual benefit at age 65 with the salary rate function as given in column 7 of Table 3.3 and annual actual salary at age 40 to be Rs 500000. We have $b_1(65) = 311467.50$ and $b_2(65) = 288165.10$, where $b_1(65)$ is the $d \times$ the salary at 65 and $b_2(65)$ is $d\times$ average of final five years salary, where $d = 0.2$. In the following example we obtain $^T P_t$ with $b_1(65)$ and $b_2(65)$ as the initial pension benefit.

Example 4.3.2 Suppose that the retirement age is $r = 65$, the entry age is 25, and the survival function is given by $S(x) = \exp\{-m(c^{x-25} - 1)\}$, $x \geq 25$, where, $m = B/\log_e C$. Assume that $B = 0.001$, $C = 1.098$, $l_r = 644.41$ (as in Table 4.3), and the salary rate function is as given in column 7 of Table 3.3 in Chap. 3. Suppose that $h(x) = e^{\eta(x-r)}$, $x \geq r$, $\eta = 0.045$, $\tau = 0.02$, $\delta = 0.05$.

(i) Find $^T P_t = dw_{r-1}l_r e^{(\tau+R)t}\ddot{a}_r^\eta$ for $t = 10, 20$, and 30 when $d = 0.12$.
(ii) Find $^T P_t(i) = b_i(65)l_r e^{(\tau+R)t}\ddot{a}_r^\eta$ for $t = 10, 20$, and 30, where $b_1(65) = 311467.90$ and $b_2(65) = 288165.10$.

Solution Using the values of $\ddot{a}_{65}^\eta$, d, and τ as specified, we compute $^T P_t = dw_{r-1}\ddot{a}_r^\eta l_r e^{(\tau+R)t}$ for three time points and three specified values of R. These are reported in Table 4.6. When the initial benefit is specified, dw_{r-1} is replaced by the specified benefit value. These are reported in Tables 4.7 and 4.8. The following is R code for these computations:

```
a <- 1.098   # C;
b <- 0.001   # B;
m <- b/log(a, base=exp(1));
e <- exp(1);
x <- 65:100;
p <- e^(m-m*a^(x-25))    # survival function S(x);
del <- 0.05;
eta <- 0.045;
v <- e^(-(del-eta));
```

Table 4.7 Terminal fund with initial benefit $b_1(65)$

t	$R = 0.02$	$R = 0$	$R = -0.01$
10	3181550167.00	2604832964.00	2356950334.00
20	4746315116.00	3181550167.00	2604832964.00
30	7080670113.00	3885954149.00	2878785639.00

Table 4.8 Terminal fund with initial benefit $b_2(65)$

t	$R = 0.02$	$R = 0$	$R = -0.014$
10	2943519130.00	2409949634.00	2180612604.00
20	4391214536.00	2943519130.00	2409949634.00
30	6550922298.00	3595222384.00	2663406250.00

```
ad <- v^(-65)*sum(v^x*p)/p[1];
ad    #ä^η_65;
lr <- 644.41;
d <- 0.12;
wr <- 6.79;
tau <- 0.02;
t <- c(10, 20, 30);
r1 <- 0.02;
y1 <- d*wr*ad*lr*e^((tau+r1)*t);
r2 <- 0;
y2 <- d*wr*ad*lr*e^((tau+r2)*t);
r3 <- (-0.01);
y3 <- d*wr*ad*lr*e^((tau+r3)*t);
y <- data.frame(t, y1, y2, y3);
y   #Table 4.6;
R1 <- 311467.90   # b1(65);
R2 <- 288165.10   # b2(65);
r1 <- 0.02;
y1 <- R1*ad*lr*e^((tau+r1)*t);
z1 <- R2*ad*lr*e^((tau+r1)*t);
r2 <- 0;
y2 <- R1*ad*lr*e^((tau+r2)*t);
z2 <- R2*ad*lr*e^((tau+r2)*t);
r3 <- (-0.01);
y3 <- R1*ad*lr*e^((tau+r3)*t);
z3 <- R2*ad*lr*e^((tau+r3)*t);
y <- data.frame(t, y1, y2, y3)   #Table 4.7;
z <- data.frame(t, z1, z2, z3)   #Table 4.8;
y;   z;
```

From Table 4.6, $^T P_{10} = 8322.94$ is the terminal fund for a stable population with $R = 0.02$ at time $t = 10$. It represents the expected purchasing price of the life annuities with initial payment dw_{r-1} with annual increase given by $e^{0.045}$ and inflation factor $\tau = 0.02$ for the expected number $l_r e^{0.02 \times 10} = 787.07$ individuals who retire at age 65. Thus, the employer has to deposit this much amount in the pension fund so that with the specified force of interest, this fund will be sufficient to pay the pension benefits at the given rate for all the retirees till they survive. The value of $^T P_t$ for $R = 0$ is intermediate to those for $R = 0.02$ and $R = -0.01$. As time increases, all increase at the rate e^{τ}, as expected.

(ii) In this setup, $^T P_t(i) = b_i(65) l_r e^{(\tau + R)t} \ddot{a}_r^{\eta}$, $i = 1, 2$. For the given data, $^T P_t$ corresponding to $b_1(65)$ are reported in Table 4.7 and corresponding to $b_2(65)$ are displayed in Table 4.8.

From Table 4.7 we note that the terminal fund at $t = 10$ for a stable population with $R = 0.02$ is 3181550167.00 if the initial pension payment is $b_1(65) = 311467.90$ and increases annually at the rate e^{η}. That is, the expected amount 3181550167.00 needs to be deposited in the pension fund to support the pension benefit for expected number of 787.08 members who retire at age 65. In Table 4.8, all the values of terminal fund are less than those in Table 4.7, as the initial benefit amount is smaller than that in Table 4.7.

The terminal funding method is not frequently used in practice as there is no security to the plan in case of financial collapse of the employer. So it is now well accepted that the cost of the pension plan must be recognized during the working times of the employees who are ultimately going to receive pensions, by actually funding amounts sufficient to provide completely for each employee's life annuity at the time of retirement. When pension plans are funded in this manner, the safety of pensions which are being paid to those who are already retired is assured and cannot be jeopardized by fluctuations in employment levels among active members or by the financial collapse of employer himself. In USA, Canada, and in almost every industrialized country, a private employer has to establish a pension plan which is properly funded.

We now discuss funding methods in which the employer (and the employee under contributory plan) sets aside funds on some systematic basis prior to the employee's retirement date. Thus periodic contributions are made on behalf of the group of active employees during their working years. We now proceed to discuss how to obtain the total normal cost rate for a group consisting of members of all ages at a specific time point.

Suppose that we have a group of $n(x, t)$ members of age x, $a < x < r$, at time t. Some of these may withdraw or may die or may suffer disability, and some will work till the age of retirement. Thus, those who are in service for next $r - x$ years, will retire at the end of $r - x$ years. The terminal funding cost corresponding to these retiring individuals is given by $^T P_{t+r-x}$. Its present value at time t is obtained by multiplying by $v^{r-x} = e^{-\delta(r-x)}$. Thus, $e^{-\delta(r-x)} \, ^T P_{t+r-x}$ represents the actuarial present value of the future benefits for the active members of age x at time t. In the normal cost function, this liability is recognized at an accrual rate $m(x)$. That is, the contribution to the pension fund for a group of members of age x at time t is given by $e^{-\delta(r-x)} \, ^T P_{t+r-x} m(x)$. Summing over the members of all ages between a to r, we get the contribution to be made to the pension fund at time t corresponding to all working members of all ages. Thus, the normal cost rate P_t in continuous set up is given by

$$P_t = \int_a^r e^{-\delta(r-x)} \, {}^T P_{t+r-x} m(x) \, dx.$$

P_t represents the amount to be deposited in the pension fund at time t, which will accumulate to a fund that will be sufficient to pay retirement benefits for a group of members who enter at age a at time u and retire $r - a$ years later. To elaborate more on this assertion, consider a group of participants who enter at time u. Their ultimate terminal funding cost rate will be ${}^T P_{u+r-a}$. At time t, corresponding to a group of individuals of age $x = a + t - u$, the contribution in the integral defining P_t is $e^{-\delta(r-x)} \, {}^T P_{t+r-x} m(x)$. In $r - x$ years until retirement, in view of interest, it will accumulate to

$$e^{-\delta(r-x)} \, {}^T P_{t+r-x} m(x) \times e^{\delta(r-x)} = {}^T P_{t+r-x} m(x) = {}^T P_{u+r-a} m(a + t - u).$$

Integrating these interest accumulated contributions over the entire time period, we obtain

$$\int_u^{u+r-a} {}^T P_{u+r-a} m(a + t - u) \, dt = {}^T P_{u+r-a},$$

which the required terminal funding cost rate. Thus, it is established that the normal cost rate function P_t completely funds the future pension benefit.

The normal cost rate P_t in discrete setup is defined as

$$P_t = \sum_{x=a+1}^r v^{(r-x)} \, {}^T P_{t+r-x} m(x).$$

Suppose the accrual function is given by $m_1(x) = \dfrac{S(x)v^x}{\sum_{a+1}^r S(y)v^y}$. With this accrual function, the normal cost rate P_t is expressible as follows:

$$P_t = \sum_{x=a+1}^r v^{(r-x)} \, {}^T P_{t+r-x} m_1(x)$$

$$= \frac{\sum_{x=a+1}^r v^{(r-x)} \, {}^T P_{t+r-x} S(x)v^x}{\sum_{a+1}^r S(y)v^y}$$

$$= \frac{\sum_{x=a+1}^r v^{(r-x)} \, {}^T P_{t+r-x} \frac{S(x)}{S(a)} v^{(x-a)}}{\sum_{a+1}^r \frac{S(y)}{S(a)} v^{(y-a)}}$$

$$= \frac{\sum_{x=a+1}^r v^{(r-x)} \, {}^T P_{t+r-x} \, _{x-a}p_a v^{(x-a)}}{\sum_{a+1}^r \, _{y-a}p_a v^{(y-u)}}.$$

The numerator of the last expression for P_t is interpreted as follows. We begin with a group of individuals who joins the pension plan at age a, conditional on survival to age x; the present value at age x at time t of the terminal fund for such a group is $e^{-\delta(r-x)}\, {}^T\!P_{t+r-x}$, when it is further multiplied by $e^{-\delta(x-a)}$, we get the present value at age a, at the time of entry. Thus numerator is the actuarial present value at the entry point of the retirement benefits. The denominator is simply the actuarial present value of a unit payment of the temporary life annuity immediate for the period of active service, that is, when the payment is done at the end of the year. Thus, the normal cost rate is similar to the net premium.

With accrual function $m_2(x) = \dfrac{S(x)e^{-\delta x}w_{x-1}e^{\tau(x-1)}}{\sum_{a+1}^{r} S(y)e^{-\delta y}w_{y-1}e^{\tau(y-1)}}$, P_t in the discrete setup is given by

$$
P_t = \frac{\sum_{x=a+1}^{r} e^{-\delta(r-x)}\, {}^T\!P_{t+r-x}\, {}_{x-a}p_a e^{-\delta(x-a)}w_{x-1}e^{\tau(x-1)}}{\sum_{a+1}^{r} {}_{y-a}p_a e^{-\delta(y-a)}w_{y-1}e^{\tau(y-1)}}.
$$

Here the denominator is the actuarial present value of a varying temporary life annuity immediate for the period of active service. Thus here also the normal cost rate is similar to the net premium.

Substituting the expression for ${}^T\!P_{t+r-x}$, which is given by

$$
{}^T\!P_{t+r-x} = d\, w_{r-1}e^{\tau(t+r-x)}n(r, t+r-x)\bar{a}_r^h,
$$

we get the expression for P_t in terms of all basic functions. It is given by

$$
P_t = e^{\tau t}\, d\, w_{r-1}\bar{a}_r^h \int_a^r e^{-(\delta-\tau)(r-x)}n(r, t+r-x)m(x)\,dx, \quad u \le t \le u+r-a.
$$

Here u indicates the time of entry. In discrete setup, P_t is given by

$$
P_t = e^{\tau t}\, d\, w_{r-1}\bar{a}_r^h \sum_{x=a+1}^{r} e^{-(\delta-\tau)(r-x)}n(r, t+r-x)m(x).
$$

Suppose that $n(r, t+r-x) = e^{R(t+r-x)}l_r$. Then P_t simplifies as follows:

$$
P_t = e^{\tau t}\, d\, w_{r-1}\bar{a}_r^h \sum_{x=a+1}^{r} e^{-(\delta-\tau)(r-x)}n(r, t+r-x)m(x)
$$

$$
= e^{\tau t}\, d\, w_{r-1}\bar{a}_r^h \sum_{x=a+1}^{r} e^{-(\delta-\tau)(r-x)}e^{R(t+r-x)}l_r m(x)
$$

$$
= e^{(\tau+R)t-(\delta-\tau-R)r}\, d\, w_{r-1}l_r\bar{a}_r^h \sum_{x=a+1}^{r} e^{(\delta-\tau-R)x}m(x).
$$

Different formulas for $m(x)$ produces different expressions for the total normal cost rate.

We have discussed the concept of accrued actuarial liability for an individual of age x. We now discuss it for a group of active lives at time t. It is denoted by $(aV)_t$ and is defined as,

$$(aV)_t = \int_a^r e^{-\delta(r-x)}\, {}^TP_{t+r-x} M(x)\,dx.$$

The definition is similar to that for an individual set up. Substituting the expression for ${}^TP_{t+r-x}$, we get the expression for $(aV)_t$ in terms of all basic functions. Thus,

$$(aV)_t = e^{\tau t}\, dw_{r-1}\bar{a}_r^h \int_a^r e^{-(\delta-\tau)(r-x)} n(r, t+r-x) M(x)\,dx, \quad u \le t \le u+r-a,$$

where u indicates the time of entry. In discrete setup, $(aV)_t$ is given by

$$(aV)_t = e^{\tau t}\, dw_{r-1}\bar{a}_r^h \sum_{x=a+1}^r e^{-(\delta-\tau)(r-x)} n(r, t+r-x) M(x).$$

Suppose that $n(r, t+r-x) = e^{R(t+r-x)} l_r$. Then $(aV)_t$ simplifies as follows:

$$(aV)_t = e^{\tau t}\, dw_{r-1}\bar{a}_r^h \sum_{x=a+1}^r e^{-(\delta-\tau)(r-x)} e^{R(t+r-x)} l_r M(x)$$

$$= e^{(\tau+R)t-(\delta-\tau-R)r}\, dw_{r-1}l_r\bar{a}_r^h \sum_{x=a+1}^r e^{(\delta-\tau-R)x} M(x).$$

It is to be noted that the expression of $(aV)_t$ is similar to that for P_t, with the only change that $m(x)$ in P_t is replaced by $M(x)$ in $(aV)_t$. the following example illustrates how P_t, $(aV)_t$ and ${}^TP_t$ are related.

Example 4.3.3 Prove that $P_t + \delta(aV)_t = {}^TP_t + \frac{d}{dt}(aV)_t$. Interpret the result.

Solution The normal cost rate P_t, given by $P_t = \int_a^r e^{-\delta(r-x)}\, {}^TP_{t+r-x} m(x)\,dx$, can be rewritten as $P_t = \int_a^r e^{-\delta(r-x)}\, {}^TP_{t+r-x} \frac{dM(x)}{dx}$. Integrating by parts, we get

$$P_t = \left[e^{-\delta(r-x)}\, {}^TP_{t+r-x} M(x) \right]_{x=a}^{x=r} - \delta \int_a^r e^{-\delta(r-x)}\, {}^TP_{t+r-x} M(x)\,dx$$

$$= -\int_a^r M(x)e^{-\delta(r-x)} \frac{d}{dx}\left({}^TP_{t+r-x}\right)dx$$

$$= {}^TP_t - \delta(aV)_t + \int_a^r M(x)e^{-\delta(r-x)} \frac{d}{dt}\left({}^TP_{t+r-x}\right)dx$$

$$= {}^TP_t - \delta(aV)_t + \frac{d}{dt}(aV)_t.$$

The third equality in the above expression follows from the fact that

$$\frac{d}{dx}\left({}^{T}P_{t+r-x}\right) = \frac{d}{dt}\left({}^{T}P_{t+r-x}\right)\frac{d}{dx}(t+r-x) = -\frac{d}{dx}\left({}^{T}P_{t+r-x}\right).$$

The expression for P_t gives the identity $P_t + \delta(aV)_t = {}^{T}P_t + \frac{d}{dt}(aV)_t$. It can be interpreted from the viewpoint of compound interest theory. The accrued actuarial liability, $(aV)_t$, is a fund into which normal costs, at rate P_t, are paid and from which terminal funding costs, at rate ${}^{T}P_t$, are transferred when active members retire. The left-hand side of the identity is the income rate to the fund from normal costs and interest. The right-hand side represents the allocation of this income rate to the terminal funding rate and the rate of change in the fund size.

The concept of accrued liability, which is the ideal fund balance, resembles to a net level premium reserve, and the normal cost is like a net premium. As in individual setup, $(aV)_t$ defined above is similar to the retrospective reserve. In life insurance, according to the prospective approach, the reserve at any time is defined as the actuarial present value of the future benefits to be paid less the actuarial present value of the future premiums to be received. We have already seen that for the individual setup, both prospective and retrospective reserves are the same. In the following we prove that the actuarial liability at time t for a group of active members is given by the actuarial present value of future pensions for active members less the actuarial present value of future normal costs. That is, two approaches lead to the same value of liability. Toward it, we define some functions as follows.

Let $(Pa)_t$ denote the actuarial present value of future normal costs. We obtain the expression for it as follows. We have already noted that $n(x, t)$ members of age x at time t will have a terminal funding cost of ${}^{T}P_{t+r-x}$ when they retire $r - x$ years later. As these members pass from age y to $y + dy$, $x \le y < r$, the normal cost $e^{-\delta(r-y)}\,{}^{T}P_{t+r-x}m(y)\,dy$ will be payable. The present value of this normal cost is

$$e^{-\delta(r-y)}\,{}^{T}P_{t+r-x}m(y)\,dy \times e^{-\delta(y-x)} = e^{-\delta(r-x)}\,{}^{T}P_{t+r-x}m(y)\,dy.$$

Then $(Pa)_t$, the actuarial present value of future normal costs for all active members at time t, is given by

$$(Pa)_t = \int_a^r e^{-\delta(r-x)}\,{}^{T}P_{t+r-x}\left\{\int_x^r m(y)\,dy\right\}dx$$

$$= \int_a^r e^{-\delta(r-x)}e^{\tau(t+r-x)}\,dw_{r-1}n(r, t+r-x)\bar{a}_r^h\left\{\int_x^r m(y)\,dy\right\}dx$$

$$= dw_{r-1}\bar{a}_r^h e^{\tau t}e^{-(\delta-\tau)r}\int_a^r e^{-(\tau-\delta)x}n(r, t+r-x)\{1-M(x)\}\,dx,$$

where $\int_x^r m(y)\,dy = 1 - M(x)$. In discrete setup, $(Pa)_t$ is given by

$$(Pa)_t = dw_{r-1}\bar{a}_r^h e^{\tau t}e^{-(\delta-\tau)r}\sum_{x=a+1}^r e^{-(\tau-\delta)x}n(r, t+r-x)\{1-M(x)\}.$$

Suppose that $n(r, t + r - x) = e^{R(t+r-x)}l_r$, that is, the population is stable with index R. Then $(Pa)_t$ simplifies as follows:

$$(Pa)_t = dw_{r-1}\bar{a}_r^h e^{\tau t} e^{-(\delta-\tau)r} \sum_{x=a+1}^{r} e^{-(\tau-\delta)x} e^{R(t+r-x)} l_r \{1 - M(x)\}$$

$$= e^{(\tau+R)t-(\delta-\tau-R)r} dw_{r-1}l_r\bar{a}_r^h \sum_{x=a+1}^{r} e^{(\delta-\tau-R)x} (1 - M(x)).$$

We use the function $(Pa)_t$ to define contribution rate in the aggregate setup in the next section.

As in the individual set up, here also we show that $(Pa)_t$ is related to $(aV)_t$, the actuarial liability at time t for active members and $(aA)_t = \int_a^r e^{-\delta(r-x)}\,{}^T P_{t+r-x}\, dx$, the actuarial present value of the future benefits for the active members. In the following we derive this relation:

$$(Pa)_t = \int_a^r e^{-\delta(r-x)}\,{}^T P_{t+r-x} \left\{ \int_x^r m(y)\, dy \right\} dx$$

$$= \int_a^r e^{-\delta(r-x)}\,{}^T P_{t+r-x} \left\{ \int_x^r dM(y) \right\} dx$$

$$= \int_a^r e^{-\delta(r-x)}\,{}^T P_{t+r-x} \{M(r) - M(x)\}\, dx$$

$$= \int_a^r e^{-\delta(r-x)}\,{}^T P_{t+r-x} \{1 - M(x)\}\, dx$$

$$= \int_a^r e^{-\delta(r-x)}\,{}^T P_{t+r-x}\, dx - \int_a^r e^{-\delta(r-x)}\,{}^T P_{t+r-x} M(x)\, dx$$

$$= (aA)_t - (aV)_t.$$

Thus we get following two identities:

$$(Pa)_t = (aA)_t - (aV)_t \quad \Leftrightarrow \quad (aV)_t = (aA)_t - (Pa)_t.$$

The left-hand side of the second identity is the actuarial liability at time t for active members and the right-hand side presents the actuarial present value of future benefits for active members less the actuarial present value of the future normal costs.

The following example illustrates computation of normal cost P_t and actuarial accrued liability $(aV)_t$ at three time points and for three types of population.

Example 4.3.4 Suppose that the retirement age is $r = 65$, the entry age is 25, and the survival function is given by, $S(x) = \exp\{-m(c^{x-25} - 1)\}$, $x \geq 25$, where, $m = B/\log_e C$. Assume that $B = 0.001$, $C = 1.098$, and $l_0 = 1000$. Suppose that $h(x) = e^{\eta(x-r)}$, $x \geq r$, $\eta = 0.045$, $\delta = 0.05$, $\tau = 0.02$, $d = 0.12$. Take $R = 0.02$, $R = 0$, and $R = -0.01$.

(i) Find P_t and $(aV)_t$ in the discrete setup for $t = 10$, 20, and 30 with $m_1(x) = \dfrac{S(x)e^{-\delta x}}{\sum_{a+1}^{r} S(y)e^{-\delta y}\,dy}$ and $m_2(x) = \dfrac{S(x)e^{-\delta x}w_{x-1}e^{\tau(x-1)}}{\sum_{a+1}^{r} S(y)e^{-\delta y}w_{y-1}e^{\tau(y-1)}}$ with dw_{r-1} as the initial annual benefit. The salary scale function is as in Table 3.3.

(ii) Find P_t when the initial annual benefit is Rs 300000/- for both accrual functions.

Solution For the given survival function and η, we have $\ddot{a}_{65}^{\eta} = 10.6254$ from Example 4.3.2. Further, $l_r = 644.41$ and $w_{r-1} = 6.79$. For the given survival function, we compute $m_1(x)$ and $m_2(x)$ and the corresponding $M_1(x)$ and $M_2(x)$. If the initial annual pension benefit after retirement at 65 is Rs 300000/-, then the dw_{r-1} is replaced by 300000. Hence, the normal cost is given by

$$P_t = 300000e^{(\tau+R)t-(\delta-\tau-R)r}l_r\bar{a}_r^h \sum_{x=a+1}^{r} e^{(\delta-\tau-R)x}m(x).$$

With these components we use the expressions of P_t and $(aV)_t$ to write the following set of R commands to compute these functions:

```
a <- 1.098   #C;
b <- 0.001   #B;
m <- b/log(a, base=exp(1));
ad <- 10.6254;
e <- exp(1);
x <- 26:65;
p <- e^(m-m*a^(x-25))   #survival function S(x);
del <- 0.05;   v <- e^(-del);
y1 <- p*v^x;
y2 <- sum(p*v^x);
y3 <- y1/y2   #m₁(x);
w1 <- cumsum(y3)   # M₁(x);
tau <- 0.02;
u <- e^(tau);
z <- read.table("D:service.txt", header=T);
z1 <- z[, 2]   #values of service function wₓ;
y4 <- p*v^x*z1*u^(x-1);
y5 <- sum(p*v^x*z1*u^(x-1));
y6 <- y4/y5   #m₂(x);
w2 <- cumsum(y6)   # M₂(x);
d <- 0.12;
t <- c(10, 20, 30);
r <- 65;
lr <- 644.41;
r1 <- 0.02;
v1 <- e^(-(del-tau-r1));
b1 <- v1^r;
b2 <- sum(y3*v1^(-x));
```

Table 4.9 Normal cost rate with $m_1(x)$	t	$R = 0.02$	$R = 0$	$R = -0.014$
	10	6417.12	3228.29	2330.76
	20	9573.22	3943.05	2575.89
	30	14281.57	4816.05	2846.80

Table 4.10 Normal cost rate with $m_2(x)$	t	$R = 0.02$	$R = 0$	$R = -0.014$
	10	6983.99	4173.52	3281.41
	20	10418.88	5097.54	3626.52
	30	15543.15	6226.15	4007.92

Table 4.11 Accrued actuarial liability with $m_1(x)$	t	$R = 0.02$	$R = 0$	$R = -0.01$
	10	197952.80	124561.80	100136.50
	20	295310.80	152140.10	110668.00
	30	440552.00	185824.40	122307.00

```
b3  <- sum(y6*v1^(-x));
v2  <- e^(tau+r1);
p1  <- d*z1[40]*ad*lr*b1*b2*v2^t
                              # P_t corresponding to m_1(x);
p2  <- d*z1[40]*ad*lr*b1*b3*v2^t
                              # P_t corresponding to m_2(x);
w3  <- sum(w1*v1^(-x));
w4  <- d*z1[40]*ad*lr*b1*w3*v2^t
                             # (aV)_t corresponding to m_1(x);
w5  <- sum(w2*v1^(-x));
w6  <- d*z1[40]*ad*lr*b1*w5*v2^t
                             # (aV)_t corresponding to m_2(x);
p1;   p2;   w4;   w6;
```

In the last 12 commands we change value of $r1$ to 0 to get objects p3 and p4 similar to p1 and p2, respectively, for P_t and objects w7 and w8 similar to w4 and w6 for $(aV)_t$. Further, changing value of $r1$ to -0.01, we get the objects p5 and p6 similar to p1 and p2, respectively, for P_t and objects w9 and w10 similar to w4 and w6 for $(aV)_t$, using data.frame function. The values of P_t corresponding to $m_1(x)$ and combined and are reported in Table 4.9, and those corresponding to $m_2(x)$ are reported in Table 4.10.

The values of $(aV)_t$ corresponding to $m_1(x)$ are reported in Table 4.11, and those corresponding to $m_2(x)$ are reported in Table 4.12.

Table 4.12 Accrued actuarial liability with $m_2(x)$	t	$R = 0.02$	$R = 0$	$R = -0.014$
	10	141549.30	93524.66	76842.47
	20	211166.70	114231.28	84924.06
	30	315023.70	139522.40	93855.61

Table 4.13 Normal cost with initial benefit 300000 with $m_1(x)$	t	$R = 0.024$	$R = 0$	$R = -0.014$
	10	2362710247.00	1188620539.00	858159368.00
	20	3524749500.00	1451784405.00	948412777.00
	30	5258308357.00	1773213476.00	1048158219.00

Table 4.14 Normal cost with initial benefit 300000 with $m_2(x)$	t	$R = 0.02$	$R = 0$	$R = -0.014$
	10	2571423240.00	1536640396.00	1208177301.00
	20	3836112698.00	1876856818.00	1335242417.00
	30	5722807666.00	2292398094.00	1475671088.00

With initial annual benefit Rs 300000/-, d*z1[40] in p1 and p2 will be replaced by 300000. Similar changes will be done in p3, p4, p5, and p6.

The normal costs, with initial annual benefit Rs 300000/-, corresponding to $m_1(x)$, are reported in Table 4.13, and those corresponding to $m_2(x)$ are reported in Table 4.14.

P_t is termed as the normal cost of the pension plan because it is the cost of keeping the pension fund at the desired level. If the underlying mortality and interest assumptions are satisfied, then the fund assets equal the accrued liability. Thus, the normal cost represents the cost under normal circumstances. The normal cost is not a proper reflection of the full cost of the plan except in this ideal setting. In real life the actual experience is not exactly equal to the mortality, and interest assumptions and hence the fund balance are not equal to the accrued liability. For example, the plan may have experienced good fortune relative to assumptions over the years, and then the assets in the fund may be in excess of the accrued liability. On the other hand, the bad experience will produce an accrued liability in excess of the assets. Therefore, although the central component of the pension cost is the normal cost, there needs adjustment in the cost to allow for these variations from the ideal. In other words, the normal cost is the expected value of the underlying random cost, and the actual realization of the random variable will be scattered around this center.

In the following section we study a method to compute the normal cost which takes into account the difference between the fund generated by the contributions and the fund needed to meet the obligations. This method is known as the actuarial cost method. In this method, the contribution rate is not determined on individual basis but on a collective basis, and it is a function of P_t, $(Pa)_t$, and $(aV)_t$ defined in this section.

4.4 Aggregate Actuarial Cost Method

The rate of contribution in aggregate or group actuarial cost method is defined in terms of the following three functions. Let $(aF)_t$ denote the fund allocated to active members at time t, $(aC)_t$ denote the contribution rate, and $(aV)_t$ denote the accrued actuarial liability at time t with respect to active participants. Then $(aU)_t$, the unfunded actuarial accrued liability at time t with respect to active participants, is given by

$$(aU)_t = (aV)_t - (aF)_t.$$

The unfunded actuarial accrued liability reflects the changes in the fund due to gains or losses as a consequence of changes in assumed mortality or retirement pattern and the rate of investment returns. In this approach some part of the unfunded actuarial accrued liability is added to the normal cost rate. Suppose that the normal cost rate P_t is determined by the accrued benefit cost methods, discussed in the previous section. The contribution rate $(aC)_t$ in the aggregate actuarial cost method is obtained by adding to the normal cost P_t some fraction of the unfunded liability and is given by

$$(aC)_t = P_t + \lambda(t)(aU)_t,$$

where $\lambda(t)$ is a fraction and is known as the process of amortizing, that is, paying off or paying back $(aU)_t$. Thus, the contribution rate depends on the magnitude of unfunded liability taking care of gains or losses resulting from the deviation from underlying assumptions. It is to be noted that with $\lambda(t) = 0$, the contribution rate in this approach is the same as the normal cost rate in the accrued benefit cost method.

As an illustration, suppose that $\lambda(t)$ is given by

$$\lambda(t) = \frac{P_t}{(Pa)_t},$$

where $(Pa)_t$ is the actuarial present value of future normal costs. Thus, $\lambda(t) = \frac{P_t}{(Pa)_t}$ represents the part of the actuarial present value of the future normal costs recognized by the normal cost at time t. With this amortization process, $(aC)_t = P_t + \frac{P_t}{(Pa)_t}(aU)_t$. The addition to the normal cost is interpreted as follows. If the actuarial present value of future normal costs, $(Pa)_t$, corresponds to the unfunded liability $(aU)_t$, then a fraction of it given by $\frac{P_t}{(Pa)_t}(aU)_t$ corresponds to the normal cost at time t. So to find the contribution rate at time t which takes into account the unfunded liability, we have to add $\frac{P_t(aU)_t}{(Pa)_t}$ to the normal cost P_t.

Suppose that $(Pa)_t = k(t)P_t$. We know that $(Pa)_t$ is the actuarial present value of the future normal costs and P_t is the normal cost at t. The constant $k(t)$ can then be interpreted as the actuarial present value of unit temporary annuity such that this temporary annuity, from t to the time of retirement, with a level income rate P_t is equal to the actuarial present value of the future normal costs $(Pa)_t$ for the active members at time t. Hence the constant $k(t)$ is denoted by $\bar{a}_{P_t}$. It is to be noted that the annuity function notation "a" is used because the nature of $k(t)$ is similar to that of annuity function. The suffix P_t is used to indicate its relation to P_t. Thus the amortization process is given by

$$\lambda(t) = \frac{1}{\bar{a}_{P_t}}, \quad \text{where } \bar{a}_{P_t} = \frac{(Pa)_t}{P_t}.$$

$\bar{a}_{P_t}$ in terms of all basic functions is given by

$$\begin{aligned}
\bar{a}_{P_t} &= \frac{(Pa)_t}{P_t} \\
&= \frac{dw_{r-1}\bar{a}_r^h e^{\tau t} e^{-(\delta-\tau)r} \int_a^r e^{-(\tau-\delta)x} n(r,t+r-x)\{1-M(x)\}\,dx}{e^{\tau t}\,dw_{r-1}\bar{a}_r^h e^{-(\delta-\tau)r} \int_a^r e^{-(\tau-\delta)x} n(r,t+r-x)m(x)\,dx} \\
&= \frac{\int_a^r e^{-(\tau-\delta)x} n(r,t+r-x)\{1-M(x)\}\,dx}{\int_a^r e^{-(\tau-\delta)x} n(r,t+r-x)m(x)\,dx}.
\end{aligned}$$

In discrete setup with $n(x,t) = e^{Rt}l_x$, we have derived that

$$(Pa)_t = e^{(\tau+R)t-(\delta-\tau-R)r}\,dw_{r-1}l_r\bar{a}_r^h \sum_{x=a+1}^r e^{(\delta-\tau-R)x}\left(1-M(x)\right)$$

and

$$P_t = e^{(\tau+R)t-(\delta-\tau-R)r}\,dw_{r-1}l_r\bar{a}_r^h \sum_{x=a+1}^r e^{(\delta-\tau-R)x}m(x).$$

Hence, in this setup,

$$\bar{a}_{P_t} = \frac{(Pa)_t}{P_t} = \frac{\sum_{x=a+1}^r e^{(\delta-\tau-R)x}(1-M(x))}{\sum_{x=a+1}^r e^{(\delta-\tau-R)x}m(x)}.$$

It is to be noted that for the stable population, $\bar{a}_{P_t}$ is free from t. We use this expression to compute $\bar{a}_{P_t}$ for stationary and stable populations at various time points in the following example for two accrual functions.

Example 4.4.1 Suppose that the retirement age is $r = 65$, the entry age is $a = 25$, and the survival function is given by $S(x) = \exp\{-m(c^{x-25}-1)\}$, $x \geq 25$, where $m = B/\log_e C$. Assume that $B = 0.001$ and $C = 1.098$. Suppose that $\delta = 0.05$ and $\tau = 0.02$. Suppose that $R = 0.02$, $R = 0$, and $R = -0.01$. Find $\bar{a}_{P_t}$ and $\lambda(t) = 1/\bar{a}_{P_t}$ in the discrete setup with $m_1(x) = \frac{S(x)e^{-\delta x}}{\sum_{a+1}^r S(y)e^{-\delta y}}$ and $m_2(x) = \frac{S(x)e^{-\delta x}w_{x-1}e^{\tau(x-1)}}{\sum_{a+1}^r S(y)e^{-\delta y}w_{y-1}e^{\tau(y-1)}}$.

Solution To compute $\bar{a}_{P_t}$, we need to find $M(x)$ for each x. We compute it as $M(x) = \sum_{y=a+1}^{x} m(y)$. The following is a set of R commands to find the values of $\bar{a}_{P_t}$ and hence of $\lambda(t)$;

```
a <- 1.098    #C;
b <- 0.001    #B;
m <- b/log(a, base=exp(1));
e <- exp(1);
x <- 26:65;
p <- e^(m-m*a^(x-25))   #survival function S(x);
del <- 0.05;
v <- e^(-del);
z <- read.table("D:service.txt", header=T);
z1 <- z[, 2]   #values of service function w_x;
y1 <- p*v^x;
y2 <- sum(p*v^x);
y3 <- y1/y2   #m_1(x);
y4 <- cumsum(y3)   #M_1(x);
tau <- 0.02;
u <- e^(tau);
y5 <- p*v^x*z1*u^(x-1);
y6 <- sum(p*v^x*z1*u^(x-1));
y7 <- y5/y6   #m_2(x);
y8 <- cumsum(y7)   #M_2(x);
r <- c(0.02, 0, -0.01);
v1 <- e^(del-tau-r[1]);
y9 <- sum(v1^x*(1-y4))/sum(v1^x*y3);
y10 <- sum(v1^x*(1-y8))/sum(v1^x*y7);
y9; y10;
```

Changing $r[1]$ in object $v1$ to $r[2]$ and $r[3]$, we get $\bar{a}_{P_t}$ corresponding to $m_1(x)$ and $m_2(x)$ for $R = 0$ and $R = -0.01$. The values of $\bar{a}_{P_t}$ and $\lambda(t)$ corresponding to $m_1(x)$ are reported in Table 4.15, and those corresponding to $m_2(x)$ are reported in Table 4.16.

From Table 4.15 we note that 8.24 % of unfunded liability is added to the normal cast for a stable population with $R = 0.02$. It is to be noted that the amortization values for $m_1(x)$ are larger than those for $m_2(x)$ for all the three types of populations. Further, for both the accrual functions, these decrease as R increases. For $R > 0$, the population size is larger than that for $R = 0$, so the proportion of unfunded liability is less for $R > 0$ as compared to $R = 0$.

With the amortization process $\lambda(t) = \frac{1}{\bar{a}_{P_t}}$, the contribution rate $(aC)_t$ is given by

$$(aC)_t = P_t + \lambda(t)(aU)_t$$

$$= P_t + \frac{(aV)_t - (aF)_t}{\bar{a}_{P_t}}$$

Table 4.15 Amortization values with $m_1(x)$

R	$\bar{a}_{P_t}$	$\lambda(t)$
0.02	12.12567	0.0824697
0	11.32447	0.0883044
−0.01	10.88211	0.0918939

Table 4.16 Amortization values with $m_2(x)$

R	$\bar{a}_{P_t}$	$\lambda(t)$
0.02	19.21759	0.0520357
0	16.19638	0.0617422
−0.01	14.82828	0.0674387

$$= \frac{(Pa)_t + (aV)_t - (aF)_t}{\bar{a}_{P_t}}$$

$$= \frac{(aA)_t - (aF)_t}{\bar{a}_{P_t}}.$$

This expression of $(aC)_t$ rewritten as $(aC)_t \bar{a}_{P_t} = (aA)_t - (aF)_t$ is interpreted as follows. A temporary annuity at the rate of $(aC)_t$ is equivalent to the actuarial present value of future benefits for active members less the fund allocated for them.

The amortization process $\lambda(t) = \frac{1}{\bar{a}_{P_t}}$ has an interesting property that for large t, the allocated fund $(aF)_t$ converges to the actuarial present value of the liability $(aV)_t$, that is, $(aU)_t$ converges to 0, and, as a consequence, for large t, $(aC)_t$ is close to P_t. We prove the result in following example.

Example 4.4.2 Prove that for the amortization process $\lambda(t) = \frac{1}{\bar{a}_{P_t}}$, for large t, $(aF)_t$ converges to $(aV)_t$, provided that $\bar{a}_{P_u} < \bar{a}_{P_\infty} = 1/\delta$ for $0 < u < t$.

Solution The progress of the fund $(aF)_u$ for the active members at time u is governed by two sources of income, contributions and the interest earned, while the fund decreases by the transfer of the terminal funding cost to a fund for retired members. Mathematically, the progress of the fund is described by the differential equation

$$\frac{d}{du}(aF)_u = (aC)_u + \delta(aF)_u - {}^T P_u = P_u + \frac{(aU)_u}{\bar{a}_{P_u}} + \delta(aF)_u - {}^T P_u.$$

We have derived in Example 4.3.3 that $\frac{d}{du}(aV)_u = P_u + \delta(aV)_u - {}^T P_u$. Thus, differentiating both sides of the equation $(aU)_u = (aV)_u - (aF)_u$ with respect to u gives the rate of change of the unfunded actuarial accrued liability as in the following differential equation:

$$\frac{d}{du}(aU)_u = \frac{d}{du}(aV)_u - \frac{d}{du}(aF)_u = -\frac{(aU)_u}{\bar{a}_{P_u}} + \delta(aU)_u = -(aU)_u\left(\frac{1}{\bar{a}_{P_u}} - \delta\right).$$

To solve this differential equation, we integrate with respect to u from 0 to t and get

$$(aU)_t = (aU)_0 e^{-\int_0^t (\frac{1}{\bar{a}_{P_u}} - \delta)\,du} \quad \Leftrightarrow$$

$$(aF)_t = (aV)_t - \left[(aV)_0 - (aF)_0\right]e^{-\int_0^t (\frac{1}{\bar{a}_{P_u}} - \delta)\,du}.$$

If $\bar{a}_{P_u} < \bar{a}_{P_\infty} = 1/\delta$ for $0 < u < t$, then as $t \to \infty$, the factor $e^{-\int_0^t (\frac{1}{\bar{a}_{P_u}} - \delta)\,du} \to 0$. Consequently, $(aF)_t \to (aV)_t$ as $t \to \infty$, and the result is proved.

It is to be noted that the accrual function is implicitly involved in computation of contribution rate in aggregate actuarial cost method, via P_t and $(aV)_t$. Different accrual functions produce a different pattern of contributions and a different ultimate fund. Thus, while referring to an aggregate actuarial cost method, it is necessary to specify the accrual function used to find P_t and $(aV)_t$. The aggregate method with entry-age accrual function, proportional to the salary scale, is particularly important in practice. In the following example we obtain the expression for $\bar{a}_{P_t}$ if the accrual function $m(x)$ is as defined in the entry-age normal actuarial cost method and is given by

$$m(x) = \frac{S(x)e^{-\delta x}e^{\tau x}w_x}{\int_a^r S(y)e^{-\delta y}e^{\tau y}w_y\,dy}.$$

Example 4.4.3 Suppose that the accrual function $m(x)$ is given by

$$m(x) = \frac{S(x)e^{-\delta x}e^{\tau x}w_x}{\int_a^r S(y)e^{-\delta y}e^{\tau y}w_y\,dy}.$$

(i) Find $\bar{a}_{P_t}$.
(ii) Show that $\bar{a}_{P_t} = \bar{a}_{W_t}$, where $\bar{a}_{W_t} = \frac{(Wa)_t}{W_t}$. The function W_t, the total payment rate at time t, is as defined Sect. 4.3, and $(Wa)_t$ is defined as

$$(Wa)_t = \int_a^r n(x,t)\left\{\int_x^r e^{-\delta(y-x)}e^{\tau(t+y-x)}\frac{S(y)}{S(x)}w_y\,dy\right\}dx.$$

Solution (i) Substituting the expression for $m(x)$ into the formula for $\bar{a}_{P_t}$, we get

$$\bar{a}_{P_t} = \frac{(Pa)_t}{P_t}$$

$$= \frac{\int_a^r e^{-(\tau-\delta)x}n(r,t+r-x)\{1 - M(x)\}\,dx}{\int_a^r e^{-(t-\delta)x}n(r,t+r-x)m(x)\,dx}$$

$$= \frac{\int_a^r e^{-\delta(r-x)+\tau(r-x)} n(r,t+r-x)\{\int_x^r e^{-(\delta-\tau)y} S(y) w_y \, dy\} dx}{\int_a^r e^{-\delta(r-x)+\tau(r-x)} n(r,t+r-x) e^{-\delta x+\tau x} S(x) w_x \, dx}$$

$$= \frac{\int_a^r e^{-(\tau-\delta)x} n(r,t+r-x)\{\int_x^r e^{-(\delta-\tau)y} S(y) w_y \, dy\} dx}{\int_a^r n(r,t+r-x) S(x) w_x \, dx}.$$

(ii) We rewrite $\bar{a}_{P_t}$ to show that $\bar{a}_{P_t} = \bar{a}_{W_t}$. We use the fact that

$$n(x,t) \,_{r-x}p_x = n(r,t+r-x) \quad \Leftrightarrow \quad S(r)n(x,t) = n(r,t+r-x)S(x).$$

With this change, the last step in the derivation of $\bar{a}_{P_t}$ can be rewritten as follows:

$$\bar{a}_{P_t} = \frac{\int_a^r e^{-(\tau-\delta)x} n(x,t) S(r)/S(x)\{\int_x^r e^{-(\delta-\tau)y} S(y) w_y \, dy\} dx}{\int_a^r n(x,t) S(r) w_x \, dx}$$

$$= \frac{\int_a^r n(x,t)\{\int_x^r e^{-\delta(y-x)} e^{\tau(y-x)} \frac{S(y)}{S(x)} w_y \, dy\} dx}{\int_a^r n(x,t) w_x \, dx}$$

$$= \frac{\int_a^r n(x,t)\{\int_x^r e^{-\delta(y-x)} e^{\tau(t+y-x)} \frac{S(y)}{S(x)} w_y \, dy\} dx}{\int_a^r n(x,t) e^{\tau t} w_x \, dx}$$

$$= \frac{(Wa)_t}{W_t},$$

where the function W_t, the total payment rate at time t, is as defined Sect. 4.3, and $(Wa)_t$ is defined as $(Wa)_t = \int_a^r n(x,t)\{\int_x^r e^{-\delta(y-x)} e^{\tau(t+y-x)} \frac{S(y)}{S(x)} w_y \, dy\} dx$. Thus, for the entry-age normal actuarial cost method, $\bar{a}_{P_t} = \bar{a}_{W_t}$.

To interpret $\bar{a}_{W_t}$, we rewrite $(Wa)_t$ as follows:

$$(Wa)_t = \int_a^r n(x,t)\left\{\int_x^r e^{-\delta(y-x)} e^{\tau(t+y-x)} \frac{S(y)}{S(x)} w_y \, dy\right\} dx$$

$$= \int_a^r n(x,t)\left\{\int_x^r e^{-\delta(y-x)} \,_{y-x}p_x e^{\tau(t+y-x)} w_y \, dy\right\} dx.$$

Suppose that the individual of age x at time t survives to age $y > x$, at time $t+y-x$, with probability $_{y-x}p_x$; then the rate of salary is $w_y e^{\tau(t+y-x)}$, and multiplying it by $e^{-\delta(y-x)}$ gives the actuarial present value of the salary rate at age x at time t. Thus, the inner integral of $(Wa)_t$ represents the actuarial present value of future wages at age x of the individual of age $y > x$; summing over all ages $y > x$ gives the actuarial present value at time t of the future wages till the time of retirement. With these functions $(Wa)_t$ and W_t, a function $\bar{a}_{W_t}$ defined as

$$\bar{a}_{W_t} = \frac{(Wa)_t}{W_t}$$

is interpreted as an average annuity value for the future wages of the active lives. It is to be noted that $\bar{a}_{W_t}$ is an average annuity value for the future wages of the active lives, while $\bar{a}_{P_t}$ is an average annuity value for the future normal costs of the active lives. In practice, the amortization process defined as $\lambda(t) = 1/\bar{a}_{W_t} = W_t/(Wa)_t$ is commonly used.

The main aim of any funding method is to find the amount of contribution rate at time t. In the following we find the expression for $(aC)_t$ in terms of basic functions and corresponding to specified amortization process. The contribution rate is given by $(aC)_t = P_t + \lambda(t)(aU)_t$. To find $(aC)_t$, we must know $(aU)_t$. As in Example 4.4.2, it can be shown that

$$\frac{d}{du}(aU)_u = -\lambda(u)(aU)_u + \delta(aU)_u = -(aU)_u\big(\lambda(u) - \delta\big)$$

$$\Leftrightarrow \quad (aU)_t = (aU)_0 e^{-\int_0^t (\lambda(u) - \delta)\,du},$$

where $(aU)_0 = (aV)_0 - (aF)_0$. Thus the contribution rate $(aC)_t$ in terms of fund at time 0, the actuarial present value of future liability at time 0, and the amortization process is given by

$$(aC)_t = P_t + \lambda(t)\big[(aV)_0 - (aF)_0\big]e^{-\int_0^t (\lambda(u) - \delta)\,du}.$$

We have proved that if $n(x, t) = e^{Rt}l_x$, then $\lambda(t) = 1/\bar{a}_{P_t}$ is free from t. We denote it by λ. Then, in this case, the contribution rate $(aC)_t$ is given by

$$(aC)_t = P_t + \lambda\big[(aV)_0 - (aF)_0\big]e^{-(\lambda - \delta)t}.$$

In the following example we find the contribution rate for the discrete setup of Example 4.4.1.

Example 4.4.4 Suppose that the retirement age is $r = 65$, the entry age is $a = 25$, and the survival function is given by $S(x) = \exp\{-m(c^{x-25} - 1)\}$, $x \geq 25$, where, $m = B/\log_e C$. Assume that $B = 0.001$, $C = 1.098$, and $l_0 = 1000$. Suppose that $R = 0.02$, $R = 0$, and $R = -0.01$.

(i) Find $(aC)_t$ with amortization process $\lambda(t) = 1/\bar{a}_{P_t}$ in the discrete setup for $t = 10, 20$ and 30, with $m_1(x) = \dfrac{S(x)e^{-\delta x}}{\sum_{a+1}^{r} S(y)e^{-\delta y}}$ and $m_2(x) = \dfrac{S(x)e^{-\delta x}w_{x-1}e^{\tau(x-1)}}{\sum_{a+1}^{r} S(y)e^{-\delta y}w_{y-1}e^{\tau(y-1)}}$.
   Assume that $(aU)_0 = 10000$.
(ii) Also find the contribution to P_t to compensate for unfunded liability.

Solution For the given setup, we have computed the values of P_t in Example 4.3.4 and of $\lambda(t) = \lambda = 1/\bar{a}_{P_t}$ in Example 4.4.1. Suppose that the values of P_t corresponding to $m_1(x)$ as computed in Example 4.3.4 are stored in file pt1.txt and the values of P_t corresponding to $m_2(x)$ are stored in file pt2.txt. Further, suppose that the values of $\lambda(t) = 1/\bar{a}_{P_t} = \lambda$ corresponding to $m_1(x)$ as computed in Example 4.4.1 are stored in object u1 for $R = 0.02$, $R = 0$, and $R = -0.01$, and those corresponding to $m_2(x)$ are stored in object u2. We use these values to find

Table 4.17 Contribution rate for $m_1(x)$

t	$R = 0.02$	$R = 0$	$R = -0.01$
10	7013.17	3830.34	2935.19
20	10004.01	4353.51	2973.45
30	14592.92	5095.90	3108.29

Table 4.18 Contribution rate for $m_2(x)$

t	$R = 0.02$	$R = 0$	$R = -0.014$
10	7493.86	4722.53	3847.88
20	10918.48	5585.74	4102.33
30	16032.68	6660.26	4407.60

the values of $(aC)_t$. In this setup, $(aC)_t$ is given by

$$(aC)_t = P_t + \lambda(aU)_0 e^{-(\lambda-\delta)t}.$$

The following set of R commands compute $(aC)_t$ for specified values of t and R. The values of $(aC)_t$ corresponding to $m_1(x)$ are reported in Table 4.17, and those corresponding to $m_2(x)$ are reported in Table 4.18.

```
z1 <- read.table("D:pt1.txt");
u1 <- c(0.0824697, 0.0883044, 0.0918939);
z2 <- read.table("D:pt2.txt");
u2 <- c(0.0520357, 0.0617422, 0.0674387);
t <- c(10, 20, 30);
u0 <- 10000;
del <- 0.05;
e <- exp(1);   y1 <- z1[, 1]+u1[1]*u0*e^((del-u1[1])*t);
y2 <- z1[, 2]+u1[2]*u0*e^((del-u1[2])*t);
y3 <- z1[, 3]+u1[3]*u0*e^((del-u1[3])*t);
y4 <- data.frame(t, y1, y2, y3);
y4  #Table 4.17;
y5 <- z2[, 1]+u2[1]*u0*e^((del-u2[1])*t);
y6 <- z2[, 2]+u2[2]*u0*e^((del-u2[2])*t);
y7 <- z2[, 3]+u2[3]*u0*e^((del-u2[3])*t);
y8 <- data.frame(t, y5, y6, y7);
y8  #Table 4.18;
y9 <- data.frame(y1, y2, y3);
y10 <- y9-z1;
y10  #Table 4.19;
y11 <- data.frame(y5, y6, y7);
y12 <- y11-z2;
y12  #Table 4.20;
```

Table 4.19 Compensation for unfunded liability for $m_1(x)$	t	$R = 0.02$	$R = 0$	$R = -0.01$
	10	596.05	602.04	604.43
	20	430.79	410.46	397.56
	30	311.35	279.85	261.49

Table 4.20 Compensation for unfunded liability for $m_2(x)$	t	$R = 0.02$	$R = 0$	$R = -0.01$
	10	509.87	549.02	566.47
	20	499.60	488.19	475.82
	30	489.53	434.11	399.67

From Tables 4.17 and 4.18 we note that the rate of contribution corresponding to $m_2(x)$ is higher than that for $m_1(x)$. These increase as time t increases in all the cases.

Compensation for the unfunded liability, for three values of t and three types of population, is obtained just by subtracting elements of z1 from the corresponding entries in Table 4.17 for the accrual function $m_1(x)$, and similarly, by subtracting elements of z2 from the corresponding entries in Table 4.18 for the accrual function $m_2(x)$. These values are reported in Tables 4.19 and 4.20.

From Tables 4.19 and 4.20 it is to be noted that, as t increases, the compensation for unfunded liability decreases for all the three values of R, supporting the result in Example 4.4.2.

4.5 Exercises

4.1 For the accrual function $m(x) = \frac{1}{r-a}$, $a \le x \le r$, find the normal constant rate $P(x)$ and the accrued actuarial liability $(aV)(x)$.

4.2 Suppose that the retirement age is $r = 65$, the entry age is 25, and the survival function is given by $S(x) = \exp\{-Ax - m(c^{x-25} - 1)\}$, $x \ge 25$, where $m = B/\log_e C$. Assume that $A = 0.0007$, $B = 0.001$, $C = 1.098$. Suppose that $h(x) = e^{\eta(x-r)}$, $\eta = 0.05$, $\delta = 0.06$.

  (i) Find $\ddot{a}_{65}^{\eta}$ corresponding to the given survival function.

  (ii) Using it, find $P(x)$ in the discrete setup when the accrual function $m_1(x) = \frac{v^x S(x)}{\sum_{y=a+1}^{r} v^y S(y)}$, $x = a+1, \dots, r$.

  (iii) Find the normal cost per annum if the initial pension benefit is Rs 250000/-.

  (iv) Also find the corresponding $(aV)(x)$.

4.3 Suppose that the retirement age is $r = 65$, the entry age is 25, and the survival function is given by $S(x) = \exp\{-Ax - m(c^{x-25} - 1)\}$, $x \ge 25$, where $m = B/\log_e C$. Assume that $A = 0.0007$, $B = 0.001$, $C = 1.098$. Suppose that $h(x) = e^{\eta(x-r)}$, $\eta = 0.05$, $\delta = 0.06$. Suppose that the accrual function is

$$m_2(x) = \frac{v^x S(x) w_{x-1} e^{\tau(x-1)}}{\sum_{y=a+1}^r v^y S(y) w_{y-1} e^{\tau(y-1)}}, \quad x = a+1, \ldots, r, \text{ where the salary rate func-}$$

tion w_x is as given in column 7 of Table 3.3 in Chap. 3, and $\tau = 0.02$.

(i) Find $\ddot{a}_{65}^{\eta}$ corresponding to the given survival function.

In the discrete setup, find the constant k, as defined in Example 4.2.4, the normal cost $P(x)$. and $(aV)(x)$ if the initial annual pension benefit is Rs 250000/-.

4.4 Prove that $(aV)(x) = (aA)(x)M(x) = (aA)(x) - (Pa)(x)$ if the accrual function $m(x)$ is proportional to $= S(x)e^{-\delta x}$.

4.5 Suppose that the retirement age is $r = 65$, the entry age is 25, and the survival function is given by, $S(x) = \exp\{-Ax - m(c^{x-25} - 1)\}$, $x \geq 25$, where $m = B/\log_e C$. Assume that $A = 0.0007$, $B = 0.001$, $C = 1.098$. The salary rate function is as given in column 7 of Table 3.3 in Chap. 3. Suppose that $h(x) = e^{\eta(x-r)}$, $\eta = 0.05$, $\tau = 0.02$, $\delta = 0.06$.

   (i) Find $\ddot{a}_{65}^{\eta}$ corresponding to the given survival function.

   (ii) Find $^T P_t = dw_{r-1} l_r e^{(\tau+R)t} \ddot{a}_r^{\eta}$ for $t = 10$, 20, and 30 when $d = 0.12$ and $R = -0.01, 0$, and 0.02.

   (iii) Find $^T P_t$ for $t = 10$, 20 and 30 and $R = -0.01, 0$ and 0.02, when initial annual pension benefit is Rs 250000/-.

   (iv) Interpret the results.

4.6 Suppose that the retirement age is $r = 65$, the entry age is 25, and the survival function is given by $S(x) = \exp\{-Ax - m(c^{x-25} - 1)\}$, $x \geq 25$, where, $m = B/\log_e C$. Assume that $A = 0.0007$, $B = 0.001$, $C = 1.098$. Suppose that $l_0 = 1000$. Suppose that $h(x) = e^{\eta(x-r)}$, $\eta = 0.045$, $\delta = 0.05$, $\tau = 0.02$, $d = 0.12$. Take $R = 0.02$, $R = 0$, and $R = -0.01$. It is given that $\ddot{a}_{65}^{\eta} = 10.6254$.

   (i) Find P_t and $(aV)_t$ in the discrete setup for $t = 10$, 20, and 30 with $m_1(x) = \frac{S(x)e^{-\delta x}}{\sum_{a+1}^r S(y)e^{-\delta y} dy}$ and $m_2(x) = \frac{S(x)e^{-\delta x} w_{x-1} e^{\tau(x-1)}}{\sum_{a+1}^r S(y)e^{-\delta y} w_{y-1} e^{\tau(y-1)}}$ with dw_{r-1} as the initial annual benefit. The salary scale function is as in Table 3.3.

   (ii) Find P_t when the initial annual benefit is Rs 300000/- for both accrual functions.

4.7 Suppose that the retirement age is $r = 65$, the entry age is $a = 25$, and the survival function is $S(x) = \exp\{-Ax - m(c^{x-25} - 1)\}$, $x \geq 25$, where $m = B/\log_e C$. Assume that $A = 0.0007$, $B = 0.001$, $C = 1.098$. Suppose that $\delta = 0.05$, $\tau = 0.02$, and $l_0 = 1000$. Suppose that $R = 0.02$, $R = 0$, and $R = -0.01$. Find $\bar{a}_{P_t}$ and $\lambda(t) = 1/\bar{a}_{P_t}$ in the discrete setup with $m_1(x) = \frac{S(x)e^{-\delta x}}{\sum_{a+1}^r S(y)e^{-\delta y}}$ and $m_2(x) = \frac{S(x)e^{-\delta x} w_{x-1} e^{\tau(x-1)}}{\sum_{a+1}^r S(y)e^{-\delta y} w_{y-1} e^{\tau(y-1)}}$.

4.8 Suppose that the retirement age is $r = 65$, the entry age is $a = 25$, and the survival function is given by $S(x) = \exp\{-Ax - m(c^{x-25} - 1)\}$, $x \geq 25$, where $m = B/\log_e C$. Assume that $A = 0.0007$, $B = 0.001$, $C = 1.098$. Suppose that $R = 0.02$, $R = 0$, and $R = -0.01$.

   (i) Find $(aC)_t$ with amortization process $\lambda(t) = 1/\bar{a}_{P_t}$ in the discrete setup for $t = 10$, 20, and 30 with $m_1(x) = \frac{S(x)e^{-\delta x}}{\sum_{a+1}^r S(y)e^{-\delta y}}$ and $m_2(x) = \frac{S(x)e^{-\delta x} w_{x-1} e^{\tau(x-1)}}{\sum_{a+1}^r S(y)e^{-\delta y} w_{y-1} e^{\tau(y-1)}}$. Assume that $(aU)_0 = 10000$.

   (ii) Also find the contribution to P_t to compensate for unfunded liability.

Chapter 5
Multi-state Transition Models for Cash Flows

5.1 Introduction

In first four chapters we have utilized the power of probability models to analyze situations involving risk, particularly to find premiums and reserves in multiple decrement models and to find the cost of pension plan. All these probability models involve transition from one state to another. For example, in a single decrement model, (x) is in one of the two states, (i) alive or (ii) dead. The only possible transition is from state alive to dead. We use the probability distribution of $T(x)$ or $K(x)$ to model the time of transition from state alive to dead. In multiple decrement model with m causes of decrement, there are $m + 1$ states for transition. We denote the state being alive as state 0 and decrement due to cause j as state j, $j = 1, 2, \ldots, m$. Then multiple decrement model describes the probabilities of transition from state 0 to state j at various time points. In this setup, transitions from j to 0 or transitions between any two states i and j, $i \neq j = 1, 2, \ldots, m$, are not possible. In multiple life models, commonly studied statuses are the joint life status or last survivor status. In this case, (x) and (y) are in one of the following four states: (i) both alive, (ii) (x) is alive, but (y) is dead, (iii) (y) is alive, but (x) is dead, and (iv) both (x) and (y) are dead. Multiple life models describe the probabilities of moving among these states at various points in time.

All the three multi-state models described above share a common characteristic: once the person leaves a state, he/she cannot return to that state. There are many instances in health insurance, disability income insurance, and vehicle insurance where the members move back and forth among states and may return to states they have previously left. For example, in disability income insurance, while modeling workers' eligibility for various employee benefits, the states considered are (i) active, (ii) temporarily disabled, (iii) permanently disabled, and (iv) inactive, which may include retirement, death, or withdrawal; these can be defined as separate states. We will discuss a model to describe the probabilities of moving among these various states, including the possibility of moving back and forth between active and temporarily disabled states several times. In vehicle insurance, in modeling insured automobile drivers' ratings by the insurer, the states considered are (i) preferred, (ii) standard, and (iii) substandard. Thus, these states describe the insured's driving record. Models are needed to describe the probabilities of transitions among these states. Sometimes a state "gone" is considered to describe that the member is no longer insured. In health insurance, in long-term care, a commonly used model is continuing care retirement communities (CCRC) model. In this model, residents may move among various states such as (i) independent living, (ii) temporarily in health center, (iii) permanently in health center, and (iv) gone.

In insurance it is of interest to see the financial impact of these transitions. Multiple state model has proved to be an appropriate model for an insurance policy in which the payment of benefits or premiums depends on being in a given state or moving between a given pair of states at a given time. In the simplest setup of single decrement model, the whole life insurance policy is issued to (x) when he is in state (i), that is, alive, premiums are payable while the insured is in state (i), and the death benefit is payable upon transition to state (ii), that is, dead. For a life annuity with single premium, single premium is paid when the contract is issued to an individual when he is in state (i). Benefits are then payable until transition to state (ii). Premiums are decided using the distribution of $T(x)$ or $K(x)$ and the force of interest. We now introduce the model for transitions among various states, when transitions among all states are possible, and see how these models are useful to study the cash flows associated with these transitions. Single and multiple decrement models are the particular cases of this general model. The most frequently used multi-state transition model is the Markov process in continuous time or Markov chain in discrete time. These stochastic processes are discussed in detail in statistics literature and are applied in a variety of areas. In the next section we introduce a simple model of Markov chain to describe the probabilities of transitions among states.

5.2 Markov Chain

Let $\{X_n, n \geq 0\}$ be a Markov chain with finite state space $S = \{1, 2, \ldots, m\}$, X_n denoting the state of the system at time n. It satisfies the Markov property given by

$$P[X_{n+1} = x_{n+1} | X_n = x_n, X_{n-1} = x_{n-1}, \ldots, X_0 = x_0] = P[X_{n+1} = x_{n+1} | X_n = x_n],$$

provided that the conditional probabilities are defined. The Markov property is usually described as "history independence," meaning that the probability distribution of the state of the system at time $n + 1$ may depend on the state at time n but does not depend on the states at times prior to n. Each X_n is a discrete random variable with set S as the set of possible values. In view of the Markov property,

$$P[X_{n+1} = x_{n+1} | X_n = x_n, X_{n-1} = x_{n-1}, \ldots, X_0 = x_0] = P[X_{n+1} = x_{n+1} | X_n = x_n],$$

the joint probabilities related to the Markov chain can be expressed in terms of the conditional distribution of X_{n+1} given $X_n = x_n$. We denote the conditional probability $P[X_{n+1} = j | X_n = i]$ by $Q_n^{(i,j)}$. Thus,

$$Q_n^{(i,j)} = P[X_{n+1} = j | X_n = i]$$

is the probability of transition from state i at time n to state j at time $n + 1$. This probability is referred to as one-step transition probability. When it depends on n, we say that the Markov chain is nonhomogeneous, and when it is free from n, then the Markov chain is known as homogeneous. For a finite state space Markov chain, transition probabilities are always presented in a matrix notation as follows. Let $Q_n = Q_n^{(i,j)}$ denote the matrix of transition probabilities from state i at time n to state j at $n + 1$, $Q_n^{(i,j)}$ being the (i, j)th element of Q_n. Thus, with state space consisting of four elements, Q_n is a 4×4 matrix given by

$$Q_n = \begin{matrix} & 1 & 2 & 3 & 4 \\ 1 \\ 2 \\ 3 \\ 4 \end{matrix} \begin{pmatrix} Q_n^{(1,1)} & Q_n^{(1,2)} & Q_n^{(1,3)} & Q_n^{(1,4)} \\ Q_n^{(2,1)} & Q_n^{(2,2)} & Q_n^{(2,3)} & Q_n^{(2,4)} \\ Q_n^{(3,1)} & Q_n^{(3,2)} & Q_n^{(3,3)} & Q_n^{(3,4)} \\ Q_n^{(4,1)} & Q_n^{(4,2)} & Q_n^{(4,3)} & Q_n^{(4,4)} \end{pmatrix}.$$

Q_n is known as a transition probability matrix. For a single decrement model, suppose that an individual is of age x at time $n = 0$. Further, suppose that 0 denotes the state that (x) is alive and 1 denotes state that (x) is dead. Then the transition probability matrix Q_n is given by

$$Q_n = \begin{matrix} & 0 & 1 \\ 0 \\ 1 \end{matrix} \begin{pmatrix} p_{x+n} & q_{x+n} \\ 0 & 1 \end{pmatrix},$$

where in usual notation, p_{x+n} denotes the probability that $(x + n)$ survives for 1 year, and q_{x+n} denotes the probability that $(x + n)$ dies in the next year.

For a multiple decrement model with m decrements, suppose that 0 denotes the state that (x) is alive, and j denotes the state that decrement occurs due to cause j, $j = 1, 2, \ldots, m$. Then the transition matrix Q_n can be specified in terms of its elements as

$$Q_n^{(0,0)} = p_{x+n}^{(\tau)}, \qquad Q_n^{(0,j)} = q_{x+n}^{(j)}, \qquad Q_n^{(j,j)} = 1, \quad \text{and}$$

$$Q_n^{(i,j)} = 0, \quad i \neq j, \; i, j = 1, 2, \ldots, m.$$

For a multiple life model with group of two individuals (x) and (y), with ages x and y at time $n = 0$, suppose that 1 denotes the state that both (x) and (y) are alive, 2 denotes the state that (x) is alive and (y) is dead, 3 denotes the state that (x) is dead and (y) is alive, and 4 denotes the state that both (x) and (y) are dead. Then transition probability matrix Q_n is given by

$$Q_n = \begin{matrix} & 1 & 2 & 3 & 4 \\ 1 \\ 2 \\ 3 \\ 4 \end{matrix} \begin{pmatrix} p_{x+n}p_{y+n} & p_{x+n}q_{y+n} & q_{x+n}p_{y+n} & q_{x+n}q_{y+n} \\ 0 & p_{x+n} & 0 & q_{x+n} \\ 0 & 0 & p_{y+n} & q_{y+n} \\ 0 & 0 & 0 & 1 \end{pmatrix}.$$

For disability income insurance, suppose that 1 denotes the state that the employee is active, 2 denotes the state that the employee is temporarily disabled, 3 denotes the state that the employee is permanently disabled, and 4 denotes the state that the employee is inactive. Then,

$$Q_n^{(3,1)} = Q_n^{(3,2)} = 0, \qquad Q_n^{(4,4)} = 1 \quad \text{and} \quad Q_n^{(4,j)} = 0 \quad \text{for } j = 1, 2, 3,$$

and other transition probabilities are obtained from observing the group of employees over a period and using statistical inference procedures to estimate the transition probabilities.

According to international actuarial notation, q denotes the failure probabilities, while p denotes the so-called success probabilities. Analogously, $Q_n^{(i,i)}$ is sometimes denoted by $P_n^{(i)}$, which is a "success probability" of remaining in state i at the next time period $n + 1$.

The transition probability matrix provides the information about the probability distribution of the state, one step in future from the given state. In practice we need to know transition probabilities for longer time periods; for example, in a single decrement model we need to know $_kp_{x+n}$ and $_kq_{x+n}$. For a nonhomogeneous Markov chain, the k-step transition probability, denoted $_kQ_n^{(i,j)}$, $k \geq 1$, is defined as

$$_kQ_n^{(i,j)} = P[X_{n+k} = j | X_n = i].$$

Thus, $_kQ_n^{(i,j)}$ denotes the conditional probability that the member is in state j after k time periods given that he is in state i at time n. It is to be noted that it is not the probability of reaching state j from state i in exactly k steps, the member may be in state j before, may have left it and returned again to j. The matrix $_kQ_n$ is used to define the corresponding $m \times m$ matrix, with $_kQ_n^{(i,j)}$ being the (i, j)th element. It is known that $_kp_{x+n} = p_{x+n} p_{x+n+1} \cdots p_{x+n+k-1}$. On similar lines it follows that

$$_kQ_n = Q_n \times Q_{n+1} \times Q_{n+2} \times \cdots \times Q_{n+k-1}, \quad k \geq 1, \quad \text{and} \quad _0Q_n = I,$$

where I is the identity matrix. Further, the probability that the member in state i at time n and remains in i for next k time periods is also given by

$$_kP_n^{(i)} = P_n^{(i)} P_{n+1}^{(i)} P_{n+2}^{(i)} \cdots P_{n+k-1}^{(i)}.$$

For a homogeneous Markov chain, Q_n does not depend on n, so we denote it by Q; then for the homogeneous Markov chain, the k-step transition probability matrix is given by

$$_kQ_n = {}_kQ = Q^k, \quad k \geq 1, \quad _0Q_n = I, \quad \text{and} \quad _kP_n^{(i)} = {}_kP^{(i)} = \left(P^{(i)}\right)^k.$$

The following examples illustrate the role of Markov models in insurance context.

Example 5.2.1 In autoinsurance, a Markov model is used for transitions among the status of drivers. Suppose that the insured drivers are classified in two classes, preferred and standard, at the end of each year. Each year 70 % of preferred are reclassified as preferred and 30 % as standard. 80 % of standard are reclassified as standard and 20 % as preferred. Find the probability that

   (i) the driver known to be classified as standard at the start of the year, will be classified as standard at the start of the third year,

  (ii) the driver is classified as standard at the start of the fourth year, given that he is in preferred state at the start of the second year,

 (iii) the driver known to be classified as standard at the start of the year will be classified as preferred at the start of the fourth year.

Solution Suppose that the preferred state is denoted by 1 and standard by 2. Since the transition probabilities do not depend on the time of transition, we model the two-state transition model as a homogeneous Markov chain with the transition probability matrix Q, given by

$$Q = \begin{array}{c} \\ 1 \\ 2 \end{array} \begin{array}{cc} 1 & 2 \\ \left(\begin{array}{cc} 0.7 & 0.3 \\ 0.2 & 0.8 \end{array} \right). \end{array}$$

We want to find (i) $P[X_3 = 2 | X_1 = 2] = {}_2 Q^{(2,2)}$ and (ii) $P[X_4 = 2 | X_2 = 1] = {}_2 Q^{(1,2)}$. We find Q^2 and hence the required transition probabilities. For (iii), we wish to find $P[X_4 = 1 | X_1 = 2] = {}_3 Q^{(2,1)}$. So we compute Q^3. The following R commands compute Q^2 and Q^3:

```
m  <- matrix(c(0.7, 0.3, 0.2, 0.8), nrow=2, ncol=2,
        byrow=TRUE)   # Q;
m1 <- m%*%m   # Q²;
m2 <- m1%*%m  # Q³;
m1;   m2;
```

We get

$$Q^2 = \begin{array}{c} \\ 1 \\ 2 \end{array} \begin{array}{cc} 1 & 2 \\ \left(\begin{array}{cc} 0.55 & 0.45 \\ 0.30 & 0.70 \end{array} \right) \end{array} \quad \text{and} \quad Q^3 = \begin{array}{c} \\ 1 \\ 2 \end{array} \begin{array}{cc} 1 & 2 \\ \left(\begin{array}{cc} 0.475 & 0.525 \\ 0.350 & 0.650 \end{array} \right). \end{array}$$

Hence, (i) ${}_2 Q^{(2,2)} = 0.70$, (ii) ${}_2 Q^{(1,2)} = 0.45$, and (iii) ${}_3 Q^{(2,1)} = 0.35$.

Example 5.2.2 Suppose that the autoinsurer classifies its policyholders according to preferred (1) and standard (2) status, starting at time 0 at the start of the first year when they are first insured, with reclassifications occurring at the start of each annual renewal of policy. The transition probability matrix Q_n, $n \geq 0$, of the corresponding nonhomogeneous Markov chain is given by

$$Q_n = \frac{1}{2}\begin{matrix}&1&2\\&\begin{pmatrix}0.6&0.4\\0.3&0.7\end{pmatrix}\end{matrix} + \left(\frac{1}{n+1}\right)\begin{matrix}&1&2\\&\begin{pmatrix}0.15&-0.15\\-0.20&0.20\end{pmatrix}\end{matrix}.$$

(i) Given that the insured is in state 1 at the start of the first year, find the probability that the insured is in state 1 at the start of the third year.

(ii) Find the probability that the insured who is in state 1 at the start of the first year transits to state 2 at the start of the fourth year.

Solution We want to find (i) $P[X_2 = 1|X_0 = 1] = {_2}Q_0^{(1,1)}$ and (ii) $P[X_3 = 2|X_0 = 1] = {_3}Q_0^{(1,2)}$. The Markov chain is nonhomogeneous; hence, ${_2}Q_0 = Q_0 \times Q_1$ and ${_3}Q_0 = Q_0 \times Q_1 \times Q_2$. We have

$$Q_0 = \frac{1}{2}\begin{matrix}&1&2\\&\begin{pmatrix}0.75&0.25\\0.10&0.90\end{pmatrix}\end{matrix}, \qquad Q_1 = \frac{1}{2}\begin{matrix}&1&2\\&\begin{pmatrix}0.675&0.325\\0.200&0.800\end{pmatrix}\end{matrix} \quad \text{and}$$

$$Q_2 = \frac{1}{2}\begin{matrix}&1&2\\&\begin{pmatrix}0.650&0.350\\0.233&0.767\end{pmatrix}\end{matrix}.$$

Hence, using matrix multiplication, we get

$${_2}Q_0 = \frac{1}{2}\begin{matrix}&1&2\\&\begin{pmatrix}0.55625&0.44375\\0.24750&0.75250\end{pmatrix}\end{matrix} \quad \text{and} \quad {_3}Q_0 = \frac{1}{2}\begin{matrix}&1&2\\&\begin{pmatrix}0.4649562&0.5350438\\0.3362075&0.6637925\end{pmatrix}\end{matrix}.$$

Hence we have (i) ${_2}Q_0^{(1,1)} = 0.55625$ and (ii) ${_3}Q_0^{(1,2)} = 0.5350438$.

We can express the higher step transition probability in terms of product of intermediate step transition probabilities. It essentially follows from the well-known Chapman–Kolmogorov equations in the theory of Markov chains. We discuss a particular expression of higher step transition probability which will be used heavily in the next section to compute the actuarial present values of cash flows. Suppose the member is in state s at time n. The probability that the member is in state i at time $n + k$ is then ${_k}Q_n^{(s,i)}$. The probability of transition from i to j in the next step is then $Q_{n+k}^{(i,j)}$. Then the product ${_k}Q_n^{(s,i)}Q_{n+k}^{(i,j)}$ is the probability that the member is in state i at time $n + k$ and in j at time $n + k + 1$, given that the member was in state s at time n. Summing over all possible intermediate states i, we get the probability of transition from state s at time n to state j after $k + 1$ steps. Mathematically,

$$\begin{aligned}{_{k+1}}Q_n^{(s,j)} &= P[X_{n+k+1} = j|X_n = s]\\ &= \sum_{i \in S} P[X_{n+k+1} = j, X_{n+k} = i|X_n = s]\end{aligned}$$

$$= \sum_{i \in S} P[X_{n+k+1} = j | X_{n+k} = i] P[X_{n+k} = i | X_n = s]$$

$$= \sum_{i \in S} {}_k Q_n^{(s,i)} Q_{n+k}^{(i,j)}.$$

Thus, the probability of transition from state s at time n to state j at time $n + k + 1$ can be obtained as the probability of transition from state s at time n, to any state i in next k steps, and the probability of transition from i at time $n + k$ to state j at time $n + k + 1$.

In the next section we discuss how these Markov models are used in insurance to analyze the financial impact of transitions through various states.

5.3 Actuarial Present Values of Cash Flows

Our main interest to study multi-state transition models is to find benefit premiums and benefit reserves when the transitions among various states of the insureds are governed by such models. In the simplest setup, that is, in single decrement model, benefit payments are made upon the failure of the alive status, and premiums are paid by the insured as long as he is in the alive status. In annuity contracts sold by the insurer, annuity payments are made to annuitants as long as they are alive. In disability income insurance models, payments are made to an employee when he is in a temporarily disabled or permanently disabled state. In auto insurance setup, an insurer is concerned about the expected claims payable and premiums collected while a driver is in a particular status. Thus, in all such models, it is of interest to study the cash flows while the member is in a particular state or upon the transition from one state to another. We discuss these in Markov chain setup.

We begin with cash flows upon transitions. Let $C^{(i,j)}$ denote a cash flow that occurs when there is a transition from state i to state j in one step. The cash flow may be from the insured to the insurer or vice-versa; as a consequence, $C^{(i,j)}$ can be positive or negative. We are interested in the actuarial present values of cash flows, so we must know the epochs of transitions. Thus, we add a prefix to $C^{(i,j)}$ to indicate the time points. Let ${}_{l+1}C^{(i,j)}$ denote the cash flow at time $l + 1$ if the member is in state i at time l and in state j at time $l + 1$. To take into account the time value of money, we denote by ${}_k v_n$ the value at time n of one unit to be paid k periods in the future at time $n + k$ when the rate of interest changes each annum. If i_n denotes the rate of interest for the period n to $n + 1$, then $v_n = (1 + i_n)^{-1}$ and ${}_k v_n = v_n v_{n+1} \cdots v_{n+k-1}$. With fixed annual effective interest rate i, ${}_k v_n = v^k$ where $v = (1 + i)^{-1}$. With these components, the actuarial present value of the cash flow is obtained as follows. Suppose that the member is in state s at time n, then the actuarial present value at time n of cash flow $C^{(i,j)}$ corresponding to future one-step transition from state i to state j, denoted $\mathrm{APV}_{s@n}(C^{(i,j)})$, is given by

$$\mathrm{APV}_{s@n}\left(C^{(i,j)}\right) = \sum_{k \geq 0} \left\{ {}_k Q_n^{(s,i)} Q_{n+k}^{(i,j)} \right\} \left\{ {}_{n+k+1}C^{(i,j)} \right\} \left\{ {}_{k+1}v_n \right\}.$$

The expression $\text{APV}_{s@n}(C^{(i,j)})$ is obtained as a sum of product of three components. The first factor corresponds to the probability of transition from state s at time n to state i at time $n + k$ and then the probability of one step transition from i at time $n + k$ to j at time $n + k + 1$ for $k = 0, 1, \ldots$. The second factor is the cash flow corresponding to transition from i at time $n + k$ to j at time $n + k + 1$; these may be 0 for some k. The third component is the discount factor to account for the time value of the cash flow, $k + 1$ time units back.

Now we proceed to discuss the actuarial present value of the cash flow when the member remains in a state for certain time period. We assume that the member is in state i at time l, and the cash flow occurs at the start of each year. Let $_l C^{(i)}$ denote the cash flow at time l if the member is in state i at time l. Suppose that the member is in state i at time n; then the actuarial present value at n of the cash flow is denoted by $\text{APV}_{i@n}(C^{(i)})$. Using arguments similar to that for $\text{APV}_{i@n}(C^{(i,j)})$, we get

$$\text{APV}_{i@n}(C^{(i)}) = \sum_{k \geq 0} {}_k Q_n^{(i,i)} {}_{n+k}C^{(i)} {}_k v_n.$$

The following example will clarify how the formula is useful to find the actuarial present values of cash flow.

Example 5.3.1 Suppose that the transitions between two states $\{1, 2\}$ are governed by the Markov chain with the following transition probability matrix:

$$Q = \begin{matrix} & 1 & 2 \\ 1 & \\ 2 & \end{matrix} \begin{pmatrix} 0.7 & 0.3 \\ 0.4 & 0.6 \end{pmatrix}.$$

Suppose that the member is in state 1 at time 1 and that there is a cash flow of 100 units for a transition from state 2 to state 1 any time in the next three periods.

(i) Suppose that the interest rate is 5 %. Compute the actuarial present value of the cash flow at time 1.
(ii) Compute the same if the interest rate is time varying, 5 % in the first year, from time n to $n + 1$, 6 % in the second, and 6.5 % in the third year.
(iii) Suppose that the member is in state 1 at time 1 and that there is a cash flow of 100 units for being in state 1 at time 1 and for the next two time points. Find the actuarial present value of cash flow at time 1.
(iv) Work out (iii) if the interest rate varies as in (ii).

Solution (i) We have to find $\text{APV}_{1@1}(C^{(2,1)})$. The Markov chain is time homogeneous, hence, it is given by

$$\text{APV}_{1@1}(C^{(2,1)}) = \sum_{k=0}^{2} \{{}_k Q_n^{(1,2)} Q_{n+k}^{(2,1)}\} \{{}_{n+k+1}C^{(2,1)}\} \{{}_{k+1} v_n\}$$

$$= \sum_{k=0}^{2} {}_{k}Q^{(1,2)}Q^{(2,1)}C^{(2,1)}v^{k+1}$$

$$= {}_{0}Q^{(1,2)}Q^{(2,1)}C^{(2,1)}v + {}_{1}Q^{(1,2)}Q^{(2,1)}C^{(2,1)}v^{2}$$

$$+ {}_{2}Q^{(1,2)}Q^{(2,1)}C^{(2,1)}v^{3}$$

$$= 0 + 100\big[0.3 \times 0.4 \times (1.05)^{-2} + \{0.7 \times 0.3 + 0.3 \times 0.6\}$$

$$\times 0.4 \times (1.05)^{-3}\big]$$

$$= 24.36.$$

(ii) With the time varying interest rates,

$$_{1}v_{n} = (1.05)^{-1} = 0.9524, \qquad {}_{2}v_{n} = {}_{1}v_{n}(1.06)^{-1} = 0.8985 \quad \text{and}$$

$$_{3}v_{n} = {}_{2}v_{n}(1.065)^{-1} = 0.8436.$$

Hence, $\mathrm{APV}_{1@1}(C^{(2,1)})$ is given by

$$\mathrm{APV}_{1@1}\big(C^{(2,1)}\big) = \sum_{k=0}^{2} {}_{k}Q^{(1,2)}Q^{(2,1)}C^{(2,1)}{}_{k+1}v_{n}$$

$$= {}_{0}Q^{(1,2)}Q^{(2,1)}C^{(2,1)}{}_{1}v_{n} + {}_{1}Q^{(1,2)}Q^{(2,1)}C^{(2,1)}{}_{2}v_{n}$$

$$+ {}_{2}Q^{(1,2)}Q^{(2,1)}C^{(2,1)}{}_{3}v_{n}$$

$$= 0 + 100\big[0.3 \times 0.4 \times 0.8985 + \{0.7 \times 0.3 + 0.3 \times 0.6\}$$

$$\times 0.4 \times 0.8436\big]$$

$$= 23.94.$$

(iii) We want to find $\mathrm{APV}_{1@1}(C^{(1)})$. Thus, with $C^{(1)} = 100$,

$$\mathrm{APV}_{1@1}(100) = \sum_{k\geq 0} {}_{k}Q^{(1,1)}100v^{k}$$

$$= 1 \times 100 \times v^{0} + 0.7 \times 100 \times v^{1} + \{0.7 \times 0.7 + 0.3 \times 0.4\} \times 100 \times v^{2}$$

$$= 100\{1 + 0.7(1.05)^{-1} + 0.61(1.05)^{-2}\}$$

$$= 222.$$

(iv) With time varying interest rates, we have

$$\mathrm{APV}_{1@1}(100) = \sum_{k\geq 0} {}_{k}Q^{(1,1)}100{}_{k}v_{n}$$

$$= 1 \times 100 \times {}_{0}v_{n} + 0.7 \times 100 \times {}_{1}v_{n} + \{0.7 \times 0.7 + 0.3 \times 0.4\}$$

$$\times 100 \times {}_{2}v_{n}$$

$$= 100\{1 + 0.7 \times 0.9524 + 0.61 \times 0.8985\}$$

$$= 221.$$

We have discussed computation of actuarial present values of cash flow for two types of scenario mainly to compute the premiums when the transitions among the states are governed by multi-state Markov models. We apply the equivalence principle to compute the benefit premium in this setup. We illustrate it with the following example.

Example 5.3.2 An insurer issues a special 3-year insurance contract to a person when the transitions among four states, 1: active, 2: disabled, 3: withdrawn, and 4: dead, are governed by the homogeneous Markov model with following transition probability matrix:

$$
Q = \begin{array}{c}
 \\
1 \\
2 \\
3 \\
4
\end{array}
\begin{pmatrix}
1 & 2 & 3 & 4 \\
0.50 & 0.25 & 0.15 & 0.10 \\
0.40 & 0.40 & 0 & 0.20 \\
0 & 0 & 1 & 0 \\
0 & 0 & 0 & 1
\end{pmatrix}.
$$

The death benefit is Rs 10000/-, payable at the end of the year of death.

(i) Suppose that the insured is in active state at the beginning of year 1 and is in disabled state at the end of year 1. Assuming the interest rate at 5 % per annum, calculate the actuarial present value of the prospective death benefit at the beginning of year 2.

(ii) Suppose that premiums are payable at the beginning of each year when the insured is active. Insureds do not pay annual premiums when they are disabled. Calculate the level annual net premium for this insurance.

Solution (i) The insurance policy is 3-year term insurance; hence the benefit is payable if death occurs during the term of three years. We want to find the actuarial present value of death benefit at the end of year 1 when the policyholder is in state 2. That is, we want to find $APV_{2@2}(C^{(1,4)}) + APV_{2@2}(C^{(2,4)})$, taking beginning of first year as $n = 1$. Here the Markov chain is homogeneous, and cash flow also is the same for all time points. Hence,

$$
\begin{aligned}
APV_{2@2}\left(C^{(1,4)}\right) &= \sum_{k=0}^{1}\{{}_kQ_n^{(2,1)}Q_{n+k}^{(1,4)}\}\{{}_{n+k+1}C^{(1,4)}\}\{{}_{k+1}v_n\} \\
&= \sum_{k=0}^{1}{}_kQ^{(2,1)}Q^{(1,4)}C^{(1,4)}v^{k+1} \\
&= {}_0Q^{(2,1)}Q^{(1,4)}C^{(1,4)}v + Q^{(2,1)}Q^{(1,4)}C^{(1,4)}v^2 \\
&= 0 \times 0.1 \times 10000 \times (1.05)^{-1} + 0.4 \times 0.1 \times 10000 \times (1.05)^{-2} \\
&= 362.81,
\end{aligned}
$$

$$\text{APV}_{2@2}\left(C^{(2,4)}\right) = \sum_{k=0}^{1}\left\{{}_kQ_n^{(2,2)}Q_{n+k}^{(2,4)}\right\}\left\{{}_{n+k+1}C^{(2,4)}\right\}\left\{{}_{k+1}v_n\right\}$$

$$= {}_0Q^{(2,2)}Q^{(2,4)}C^{(2,4)}v + Q^{(2,2)}Q^{(2,4)}C^{(2,4)}v^2$$

$$= 1 \times 0.2 \times 10000 \times (1.05)^{-1} + 0.4 \times 0.2 \times 10000 \times (1.05)^{-2}$$

$$= 2630.39.$$

Hence, the actuarial present value of the prospective death benefit at the beginning of year 2 is $\text{APV}_{2@2}(C^{(1,4)}) + \text{APV}_{2@2}(C^{(2,4)}) = 2993.20$.

(ii) To find the level annual net premium, the first step is to find the actuarial present value of the prospective death benefit at the beginning of year 1, assuming that the policyholder is active, that is, in state 1, and the second step is to find the actuarial present value of the premiums at the beginning of year 1, when the premiums are paid only when the policyholder is active. As in (i), the actuarial present value of death benefit at the beginning of year 1 when the policyholder is in state 1 is given by $\text{APV}_{1@1}(C^{(1,4)}) + \text{APV}_{1@1}(C^{(2,4)})$. We have

$$\text{APV}_{1@1}\left(C^{(1,4)}\right) = \sum_{k=0}^{2}{}_kQ^{(1,1)}Q^{(1,4)}C^{(1,4)}v^{k+1}$$

$$= {}_0Q^{(1,1)}Q^{(1,4)}C^{(1,4)}v + Q^{(1,1)}Q^{(1,4)}C^{(1,4)}v^2$$

$$+ {}_2Q^{(1,1)}Q^{(1,4)}C^{(1,4)}v^3$$

$$= 1 \times 0.1 \times 10000 \times (1.05)^{-1} + 0.5 \times 0.1 \times 10000 \times (1.05)^{-2}$$

$$+ \{0.5 \times 0.5 + 0.25 \times 0.4\} \times 0.1 \times 10000 \times (1.05)^{-3}$$

$$= 1708.24,$$

$$\text{APV}_{1@1}\left(C^{(2,4)}\right) = \sum_{k=0}^{2}{}_kQ^{(1,2)}Q^{(2,4)}C^{(2,4)}v^{k+1}$$

$$= {}_0Q^{(1,2)}Q^{(2,4)}C^{(2,4)}v + Q^{(1,2)}Q^{(2,4)}C^{(2,4)}v^2 + {}_2Q^{(1,2)}C^{(2,4)}v^3$$

$$= 0 + 0.25 \times 0.2 \times 10000 \times (1.05)^{-2}$$

$$+ \{0.5 \times 0.25 + 0.25 \times 0.4\} \times 0.2 \times 10000 \times (1.05)^{-3}$$

$$= 842.24.$$

Hence, the actuarial present value of death benefit at the beginning of year 1 when the policyholder is in state 1 is given by $\text{APV}_{1@1}(C^{(1,4)}) + \text{APV}_{1@1}(C^{(2,4)}) = 2550.48$. To find the actuarial present value of the premiums at the beginning of year 1, we use the formula for $\text{APV}_{1@1}(C^{(1)}) = \text{APV}_{1@1}(P)$, where P denotes the annual level premium. Thus,

$$\text{APV}_{1@1}(P) = \sum_{k=0}^{2} {}_{k}Q^{(1,1)}Pv^{k}$$

$$= 1 \times P \times v^{0} + 0.5 \times P \times v^{1} + \{0.5 \times 0.5 + 0.25 \times 0.4\} \times P \times v^{2}$$

$$= P\{1 + 0.5(1.05)^{-1} + 0.35(1.05)^{-2}\}$$

$$= 1.7937P.$$

With the equivalence principle, the benefit premium is obtained as

$$\text{APV}_{1@1}\big(C^{(1,4)}\big) + \text{APV}_{1@1}\big(C^{(2,4)}\big) = \text{APV}_{1@1}(P) \quad \Leftrightarrow \quad 2550.48 = 1.7937P.$$

Solving for P, we get $P = 1421.91$.

The next example illustrates the computation of cash flows in long-term care CCRC model, described in Sect. 5.1, when the transitions among the states are governed by a nonhomogeneous Markov model.

Example 5.3.3 In the CCRC model, residents move among four states, independent living: 1, temporarily in health center: 2, permanently in health center: 3, and gone: 4. Suppose that the transitions among these states are modeled by a nonhomogeneous Markov chain. The transition probability matrices at time $n \geq 0$ are specified as follows:

$$Q_0 = \begin{array}{c} \\ 1 \\ 2 \\ 3 \\ 4 \end{array}\begin{pmatrix} 0.80 & 0.10 & 0.05 & 0.05 \\ 0.20 & 0.60 & 0.10 & 0.10 \\ 0 & 0 & 0.80 & 0.20 \\ 0 & 0 & 0 & 1 \end{pmatrix}, \qquad Q_1 = \begin{array}{c} \\ 1 \\ 2 \\ 3 \\ 4 \end{array}\begin{pmatrix} 0.70 & 0.15 & 0.10 & 0.05 \\ 0.20 & 0.50 & 0.20 & 0.10 \\ 0 & 0 & 0.70 & 0.30 \\ 0 & 0 & 0 & 1 \end{pmatrix},$$

$$Q_2 = \begin{array}{c} \\ 1 \\ 2 \\ 3 \\ 4 \end{array}\begin{pmatrix} 0.60 & 0.15 & 0.15 & 0.10 \\ 0.20 & 0.40 & 0.25 & 0.15 \\ 0 & 0 & 0.60 & 0.40 \\ 0 & 0 & 0 & 1 \end{pmatrix}, \qquad Q_3 = \begin{array}{c} \\ 1 \\ 2 \\ 3 \\ 4 \end{array}\begin{pmatrix} 0.50 & 0.20 & 0.20 & 0.10 \\ 0.20 & 0.30 & 0.35 & 0.15 \\ 0 & 0 & 0.50 & 0.50 \\ 0 & 0 & 0 & 1 \end{pmatrix},$$

$$Q_4 = \begin{array}{c} \\ 1 \\ 2 \\ 3 \\ 4 \end{array}\begin{pmatrix} 0.40 & 0.20 & 0.20 & 0.20 \\ 0.10 & 0.30 & 0.30 & 0.30 \\ 0 & 0 & 0.40 & 0.60 \\ 0 & 0 & 0 & 1 \end{pmatrix}, \qquad Q_5 = \begin{array}{c} \\ 1 \\ 2 \\ 3 \\ 4 \end{array}\begin{pmatrix} 0.30 & 0.20 & 0.30 & 0.20 \\ 0.10 & 0.20 & 0.40 & 0.30 \\ 0 & 0 & 0.30 & 0.70 \\ 0 & 0 & 0 & 1 \end{pmatrix},$$

$$Q_6 = \begin{array}{c} \\ 1 \\ 2 \\ 3 \\ 4 \end{array}\begin{pmatrix} 0.20 & 0.20 & 0.30 & 0.30 \\ 0.10 & 0.10 & 0.40 & 0.40 \\ 0 & 0 & 0.20 & 0.80 \\ 0 & 0 & 0 & 1 \end{pmatrix}, \qquad Q_7 = \begin{array}{c} \\ 1 \\ 2 \\ 3 \\ 4 \end{array}\begin{pmatrix} 0.10 & 0.10 & 0.30 & 0.50 \\ 0.05 & 0.05 & 0.30 & 0.60 \\ 0 & 0 & 0.10 & 0.90 \\ 0 & 0 & 0 & 1 \end{pmatrix},$$

$$\text{and for } n \geq 8, \quad Q_n = \begin{array}{c} 1 \\ 2 \\ 3 \\ 4 \end{array}\begin{pmatrix} 0 & 0 & 0 & 1 \\ 0 & 0 & 0 & 1 \\ 0 & 0 & 0 & 1 \\ 0 & 0 & 0 & 1 \end{pmatrix}.$$

Cash flows $_{l+1}C^{(i,j)}$ occurring at time $l+1$ are as given in the following matrices $_{l+1}C$:

$$_1C = \begin{array}{c} 1 \\ 2 \\ 3 \\ 4 \end{array}\begin{pmatrix} 12 & 14 & 16 & 17 \\ 19 & 22 & 23 & 25 \\ 0 & 0 & 29 & 31 \\ 0 & 0 & 0 & 0 \end{pmatrix}, \qquad _2C = \begin{array}{c} 1 \\ 2 \\ 3 \\ 4 \end{array}\begin{pmatrix} 32 & 34 & 36 & 37 \\ 39 & 42 & 43 & 45 \\ 0 & 0 & 49 & 51 \\ 0 & 0 & 0 & 0 \end{pmatrix},$$

$$_3C = \begin{array}{c} 1 \\ 2 \\ 3 \\ 4 \end{array}\begin{pmatrix} 52 & 54 & 56 & 57 \\ 59 & 62 & 63 & 65 \\ 0 & 0 & 69 & 71 \\ 0 & 0 & 0 & 0 \end{pmatrix}, \qquad _4C = \begin{array}{c} 1 \\ 2 \\ 3 \\ 4 \end{array}\begin{pmatrix} 65 & 67 & 69 & 73 \\ 71 & 72 & 73 & 75 \\ 0 & 0 & 79 & 81 \\ 0 & 0 & 0 & 0 \end{pmatrix},$$

$$_5C = \begin{array}{c} 1 \\ 2 \\ 3 \\ 4 \end{array}\begin{pmatrix} 72 & 74 & 76 & 77 \\ 79 & 80 & 81 & 85 \\ 0 & 0 & 86 & 87 \\ 0 & 0 & 0 & 0 \end{pmatrix}, \qquad _6C = \begin{array}{c} 1 \\ 2 \\ 3 \\ 4 \end{array}\begin{pmatrix} 75 & 77 & 79 & 83 \\ 81 & 82 & 83 & 85 \\ 0 & 0 & 89 & 91 \\ 0 & 0 & 0 & 0 \end{pmatrix},$$

$$_7C = \begin{array}{c} 1 \\ 2 \\ 3 \\ 4 \end{array}\begin{pmatrix} 82 & 84 & 86 & 87 \\ 89 & 90 & 91 & 95 \\ 0 & 0 & 96 & 97 \\ 0 & 0 & 0 & 0 \end{pmatrix}, \qquad _8C = \begin{array}{c} 1 \\ 2 \\ 3 \\ 4 \end{array}\begin{pmatrix} 95 & 97 & 99 & 100 \\ 101 & 102 & 103 & 105 \\ 0 & 0 & 109 & 111 \\ 0 & 0 & 0 & 0 \end{pmatrix},$$

$$\text{and for } l \geq 8, \quad _{l+1}C = \begin{array}{c} 1 \\ 2 \\ 3 \\ 4 \end{array}\begin{pmatrix} 0 & 0 & 0 & 115 \\ 0 & 0 & 0 & 118 \\ 0 & 0 & 0 & 119 \\ 0 & 0 & 0 & 120 \end{pmatrix}.$$

(i) Suppose that a resident is in state 1 at time 0. Compute the actuarial present value of the cash flow from state 1 to state 3 or state 4, assuming the annual effective rate of interest to be 0.05.

(ii) Suppose that a resident is in state 1 at time 0. Compute the benefit premium payable at the start of the year in which the resident is in state 1 in order to finance the future cash flows of transition from state 1 to state 3.

Solution (i) The actuarial present value of cash flow at the beginning of year 1 when the resident is in state 1 is given by $\text{APV}_{1@0}(C^{(1,3)}) + \text{APV}_{1@0}(C^{(1,4)})$. Now,

$$\text{APV}_{1@0}\big(C^{(1,3)}\big) = \sum_{k=0}^{7} \big\{{}_kQ_0^{(1,1)}Q_k^{(1,3)}\big\}\big\{{}_{k+1}C^{(1,3)}\big\}\big\{v^{k+1}\big\},$$

$$\text{APV}_{1@0}\big(C^{(1,4)}\big) = \sum_{k=0}^{7} \big\{{}_kQ_0^{(1,1)}Q_k^{(1,4)}\big\}\big\{{}_{k+1}C^{(1,4)}\big\}\big\{v^{k+1}\big\}.$$

Thus, we need to calculate $_kQ_0 = Q_0 \times Q_1 \times \cdots \times Q_{k-1}$, $k \geq 1$. Using the R command for the product of matrices, we find $_kQ_0$. Using the above formulas, the actuarial present values are given by $\text{APV}_{1@0}(C^{(1,3)}) = 10.07$, $\text{APV}_{1@0}(C^{(1,4)}) = 6.96$, and $\text{APV}_{1@0}(C^{(1,3)}) + \text{APV}_{1@0}(C^{(1,4)}) = 17.04$.

(ii) To compute the benefit premium to finance the future cash flows at transition from state 1 to state 3, we need $\text{APV}_{1@0}(C^{(1,3)})$ and $\text{APV}_{1@0}(C^{(1)})$. We have $\text{APV}_{1@0}(C^{(1,3)}) = 10.07$. With $C^{(1)} = P$, $\text{APV}_{1@0}(C^{(1)})$ is given by

$$\text{APV}_{1@1}(P) = \sum_{k=0}^{7} {}_kQ^{(1,1)}Pv^k$$

$$= 2.2136P.$$

With the equivalence principle, the benefit premium is obtained as

$$\text{APV}_{1@0}\big(C^{(1,3)}\big) = \text{APV}_{1@0}(P) \quad \Leftrightarrow \quad 10.07 = 2.2136P \quad \Leftrightarrow \quad P = 4.55.$$

The following R code is used for all these computations:

```
m0 <- matrix(c(0.80, 0.10, 0.05, 0.05, 0.20, 0.60, 0.10,
        0.10, 0, 0, 0.80, 0.20, 0, 0, 0, 1), nrow=4,
        ncol=4, byrow=TRUE);
m1 <- matrix(c(0.70, 0.15, 0.10, 0.05, 0.20, 0.50, 0.20,
        0.10, 0, 0, 0.70, 0.30, 0, 0, 0, 1), nrow=4,
        ncol=4, byrow=TRUE);
m2 <- matrix(c(0.60, 0.15, 0.15, 0.10, 0.20, 0.40, 0.25,
        0.15, 0, 0, 0.60, 0.40, 0, 0, 0, 1), nrow=4,
        ncol=4, byrow=TRUE);
m3 <- matrix(c(0.50, 0.20, 0.20, 0.10, 0.20, 0.30, 0.35,
        0.15, 0, 0, 0.50, 0.50, 0, 0, 0, 1), nrow=4,
        ncol=4, byrow=TRUE);
m4 <- matrix(c(0.40, 0.20, 0.20, 0.20, 0.10, 0.30, 0.30,
        0.30, 0, 0, 0.40, 0.60, 0, 0, 0, 1), nrow=4,
        ncol=4, byrow=TRUE);
m5 <- matrix(c(0.30, 0.20, 0.30, 0.20, 0.10, 0.20, 0.40,
        0.30, 0, 0, 0.30, 0.70, 0, 0, 0, 1), nrow=4,
        ncol=4, byrow=TRUE);
m6 <- matrix(c(0.20, 0.20, 0.30, 0.30, 0.10, 0.10, 0.40,
        0.40, 0, 0, 0.20, 0.80, 0, 0, 0, 1), nrow=4,
        ncol=4, byrow=TRUE);
```

```
m7 <- matrix(c(0.10, 0.10, 0.30, 0.50, 0.05, 0.05, 0.30,
      0.60, 0, 0, 0.10, 0.90, 0, 0, 0, 1), nrow=4,
      ncol=4, byrow=TRUE);
m8 <- matrix(c(0, 0, 0, 1, 0, 0, 0, 1, 0, 0, 0, 1, 0, 0,
      0, 1), nrow=4, ncol=4, byrow=TRUE);
a1 <- m0%*%m1;
a2 <- a1%*%m2;
a3 <- a2%*%m3;
a4 <- a3%*%m4;
a5 <- a4%*%m5;
a6 <- a5%*%m6;
a7 <- a6%*%m7;
a8 <- a7%*%m8;
a9 <- a8%*%m8;
a1;  a2;  a3;  a4;  a5;  a6;  a7;  a8;
v <- (1.05)^(-1);
b1 <- 1*0.05*16*v+0.58*0.1*36*v^2+0.382*0.15*56*v^3
      +0.222*0.2*69*v^4+0.10109*0.2*76*v^5
      +0.038454*0.3*79*v^6+0.011338*0.3*86*v^7
      +0.0017007*0.3*99*v^8;
b1   # APV_{1@0}(C^{(1,3)});
b2 <- 1*0.05*17*v+0.58*0.05*37*v^2+0.382*0.1*57*v^3
      +0.222*0.1*73*v^4+0.10109*0.2*77*v^5
      +0.038454*0.2*83*v^6+0.011338*0.3*87*v^7
      +0.0017007*0.5*100*v^8;
b2   # APV_{1@0}(C^{(1,4)});
b1+b2;
b3 <- 1+0.58*v+0.382*v^2+0.222*v^3+0.10109*v^4
      +0.038454*v^5+0.011338*v^6+0.0017007*v^7;
b3   # APV_{1@0}(P);
p <- b1/b3;
p   # P;
```

We used the equivalence principle to compute the benefit premiums for Markov model; on similar lines we can compute the benefit reserve for the Markov model. We illustrate it with an example. We use the setup of Example 5.3.2.

Example 5.3.4 An insurer issues a special 3-year insurance contract to a person when the transitions among four states, 1: active, 2: disabled, 3: withdrawn, and 4: dead, are governed by the homogeneous Markov model with following transition probability matrix:

$$Q = \begin{array}{c} \\ 1 \\ 2 \\ 3 \\ 4 \end{array} \begin{array}{cccc} 1 & 2 & 3 & 4 \\ \begin{pmatrix} 0.50 & 0.25 & 0.15 & 0.10 \\ 0.4 & 0.4 & 0 & 0.2 \\ 0 & 0 & 1 & 0 \\ 0 & 0 & 0 & 1 \end{pmatrix} \end{array}.$$

The death benefit is Rs 10000/-, payable at the end of the year of death. Suppose that the insured is active at the issue of policy. Insureds do not pay annual premiums when they are disabled. Suppose that the interest rate is 5 % per annum. Calculate the benefit reserve at the beginning of year 2 and 3.

Solution By the prospective approach of reserves, the reserve at the beginning of year i, $i = 2, 3$, will be the actuarial present value of the death benefit at the beginning of year i, minus the actuarial present value of the future premiums at the beginning of year i, $i = 2, 3$. At the beginning of the second year the policyholder may be in any one of the four states. If he is in state 3 or 4, there is no future liability, so the reserve in this case is 0. Suppose that the policyholder is active or disabled at the beginning of the second year. Thus, the reserve at the beginning of year 2, denoted R_2, is given by

$$R_2 = Q^{(1,1)}\left[\text{APV}_{1@2}\left(C^{(1,4)}\right) + \text{APV}_{1@2}\left(C^{(2,4)}\right)\right]$$
$$+ Q^{(1,2)}\left[\text{APV}_{2@2}\left(C^{(1,4)}\right) + \text{APV}_{2@2}\left(C^{(2,4)}\right)\right]$$
$$- \left\{Q^{(1,1)}\text{APV}_{1@2}\left(C^{(1)}\right) + Q^{(1,2)}\text{APV}_{2@2}\left(C^{(1)}\right)\right\}.$$

In the second part of Example 5.3.2, we have calculated the premium $P = 1421.91$. We have also computed $\text{APV}_{2@2}(C^{(1,4)}) + \text{APV}_{2@2}(C^{(2,4)}) = 2993.20$ in the first part of Example 5.3.2. On similar lines we compute $\text{APV}_{1@2}(C^{(1,4)}) + \text{APV}_{1@2}(C^{(2,4)}) = 1859.51$. Actuarial present value of inflow at the beginning of the year 2 is obtained as follows. Let $C^{(1)} = P$. Then

$$Q^{(1,1)}\text{APV}_{1@2}\left(C^{(1)}\right) + Q^{(1,2)}\text{APV}_{2@2}\left(C^{(1)}\right)$$
$$= Q^{(1,1)}\sum_{k=0}^{1}{}_k Q^{(1,1)} P v^k + Q^{(1,2)}\sum_{k=0}^{1}{}_k Q^{(2,1)} P v^k$$
$$= Q^{(1,1)}\left\{P + Q^{(1,1)} P v\right\} + Q^{(1,2)}\left\{Q^{(2,1)} P v\right\}$$
$$= Q^{(1,1)} P + P v\left\{Q^{(1,1)} Q^{(1,1)} + Q^{(1,2)} Q^{(2,1)}\right\}$$
$$= Q^{(1,1)} P + {}_2 Q^{(1,1)} P v.$$

With these components, we get $R_2 = 549.56$. Similarly, the reserve at the beginning of year 3, denoted R_3, is given by

$$R_3 = {}_2 Q^{(1,1)}\left[\text{APV}_{1@3}\left(C^{(1,4)}\right) + \text{APV}_{1@3}\left(C^{(2,4)}\right)\right]$$
$$+ {}_2 Q^{(1,2)}\left[\text{APV}_{2@3}\left(C^{(1,4)}\right) + \text{APV}_{2@3}\left(C^{(2,4)}\right)\right]$$
$$- \left\{{}_2 Q^{(1,1)}\text{APV}_{1@3}\left(C^{(1)}\right) + {}_2 Q^{(1,2)}\text{APV}_{2@3}\left(C^{(1)}\right)\right\}.$$

Using the similar procedure as in Example 5.3.2, we get

$$\text{APV}_{1@3}\left(C^{(1,4)}\right) = 0.1 \times 10000 \times (1.05)^{-1} = 952.38, \qquad \text{APV}_{1@3}\left(C^{(2,4)}\right) = 0,$$
$$\text{APV}_{2@3}\left(C^{(1,4)}\right) = 0, \qquad \text{APV}_{2@3}\left(C^{(2,4)}\right) = 0.2 \times 10000 \times (1.05)^{-1} = 1904.76.$$

Further,

$$2Q^{(1,1)}\text{APV}_{1@3}(C^{(1)}) + {}_2Q^{(1,2)}\text{APV}_{2@3}(C^{(1)}) = {}_2Q^{(1,1)}P.$$

Hence,

$$R_3 = 264.23.$$

It is to be noted that $R_3 < R_2$.

Multi-state Markov models can also be applied in situations other than insurance, and the member can be a machinery or the loan contract. The following example illustrates such a situation.

Example 5.3.5 A machine can be in one of four possible states, labeled $a, b, c,$ and d. It makes transitions among these states according to Markov chain with following transition probability matrix:

$$Q = \begin{array}{c} \\ a \\ b \\ c \\ d \end{array} \begin{array}{cccc} a & b & c & d \\ \left(\begin{array}{cccc} 0.2 & 0.8 & 0 & 0 \\ 0.6 & 0 & 0.4 & 0 \\ 0.7 & 0 & 0 & 0.3 \\ 1 & 0 & 0 & 0 \end{array} \right). \end{array}$$

At time 0 the machine is in state a. A company will pay 500 units at the end of three years if the machine is in state a. Assuming the annual effective rate of interest 5 %, calculate the actuarial present value at time 0 of this payment.

Solution The actuarial present value at time 0 of the payment is given by

$$\text{APV} = {}_3Q^{(a,a)} \times 500 \times (1.05)^{-3}.$$

For the given matrix Q, we find Q^3, which gives

$$_3Q^{(a,a)} = 0.424, \quad \text{and hence} \quad \text{APV} = 183.13.$$

In the next section we briefly discuss how the actuarial present value of cash flows is computed when the transitions among states are modeled by a Markov process in continuous time, which is essentially a continuous-time Markov chain.

5.4 Markov Process Model

Let $\{X(t), t \geq 0\}$ be a Markov process with finite state space $S = \{1, 2, \ldots, m\}$, $X(t)$, denoting the state of the system at time t. It satisfies the Markov property given by

$$P\big[X(s+t) = j | X(s) = i, X(u) = x(u), 0 \leq u \leq s\big] = P\big[X(s+t) = j | X(s) = i\big],$$

provided that the conditional probabilities are defined. Thus, the future of the process, after time s, depends only at the state at time s and not on the history of the process up to time s. Each $X(t)$ is a discrete random variable with set S as the set of possible values. In view of Markov property, the probability structure depends on the transition probabilities defined as

$$p_{ij}(s, s+t) = P[X(s+t) = j | X(s) = i],$$

$$\text{with } \sum_{j=1}^{k} p_{ij}(s, s+t) = 1 \text{ for all } s, t \geq 0.$$

If $p_{ij}(s, s+t)$ depends only on t, then it is a time-homogeneous Markov process. In the following we study how the transitions among various states are modeled by Markov process in insurance set up.

Suppose $\{X(t), t \geq 0\}$ is a Markov process with finite state space $S = \{0, 1, \ldots, m\}$, with instantaneous transitions being possible between selected pairs of states. In insurance setup, these states represent different conditions for an individual or for a group of individuals. The event $X(t) = i$ means that the individual or a group of individuals is in state i at age $x + t$; thus, at time $t = 0$, that is, at the time of signing the contract, the age of the individual is assumed to be x. As an illustration, we consider the joint life and the last survivor model. Let x and y denote the ages of the husband and wife, respectively, when the annuity or the insurance policy is purchased. For $t \geq 0$, the event $X(t) = 0$ indicates that both husband and wife are alive at ages $x + t$ and $y + t$, respectively; $X(t) = 1$ indicates that husband is alive at age $x + t$ and wife died before age $y + t$; $X(t) = 2$ indicates that husband died before age $x + t$ and wife is still alive at age $y + t$; $X(t) = 3$ indicates that husband died before age $x + t$ and wife died before age $y + t$. In a permanent disability model there are three states: healthy, denoted by 0, disabled, denoted by 1, and dead, denoted by 2. Possible transitions are from 0 to 1, 0 to 2, and from 1 to 2. It is an appropriate model for a policy which provides some or all of the following benefits: (i) an annuity while permanently disabled, (ii) a lump sum on being permanently disabled, and (iii) a lump sum on death. Premiums are payable when a person is in healthy state. In this model the transition from state 1 to 0 is not possible; hence it is refereed to as permanent disability model. In a disability income insurance model, there are three states: healthy, denoted by 0, sick, denoted by 1, and dead, denoted by 2. Possible transitions are from 0 to 1, 0 to 2, 1 to 0, and from 1 to 2. Disability income insurance policy pays a benefit during period of sickness, the benefit ceases on recovery. Here also premiums are payable when a person is in healthy state. Such a model incorporates the fact that there could be several periods of sickness before death, with healthy periods in between.

We now introduce the notation for transition probabilities in Markov model in the context of insurance. For states i and j in a multi-state model and for $x, t \geq 0$, let

$$_t p_x^{ij} = P[X(x+t) = j | X(x) = i],$$

$$_t p_x^{\overline{ii}} = P[X(x+s) = i \ \forall s \in [0, t] | X(x) = i].$$

Thus, $_t p_x^{ij}$ is the probability that a life aged x in state i is in state j at age $x + t$, and $_t p_x^{\overline{ii}}$ is the probability that a life aged x in state i remains in state i throughout the period from age x to $x + t$. In this setup, the age is treated as the time parameter in the Markov process. $_t p_x^{\overline{ii}} = {_t p_x^{ii}}$ if state i cannot be reentered once it has been left. In a single decrement model, $_t p_x^{00} = {_t p_x}$, $_t p_x^{01} = {_t q_x}$, and $_t p_x^{10} = 0$, $_t p_x^{11} = 1$.

For $i \neq j$, the transition intensity function corresponding to transitions between states i and j at age x is defined as

$$\mu_x^{ij} = \lim_{h \to 0^+} \frac{_h p_x^{ij}}{h}.$$

The transition intensity function is analogous to the force of decrement or force of mortality, and hence it is also referred to as the force of transition. For $h > 0$, μ_x^{ij} can be expressed as

$$_h p_x^{ij} = h \mu_x^{ij} + o(h).$$

For small positive values of h, $_h p_x^{ij}$ is taken as $h \mu_x^{ij}$. It is further assumed that for any positive interval of time h, the probability of two or more transitions within a time period of length h is $o(h)$.

If there are $n + 1$ states, then $_t p_x^{\overline{ii}}$ can be expressed in terms of the force of transition as follows:

$$_t p_x^{\overline{ii}} = \exp\left[-\left\{\int_0^t \sum_{j=0, j \neq i}^n \mu_{x+s}^{ij} \, ds\right\}\right].$$

This expression is analogous to

$$_t p_x = \exp\left[-\left\{\int_0^t \mu_{x+s} \, ds\right\}\right]$$

in a single decrement table and

$$_t p_x^{(\tau)} = \exp\left[-\left\{\int_0^t \sum_{j=0}^n \mu_{x+s}^{(j)} \, ds\right\}\right]$$

in a multiple decrement model.

In the following example we compute certain probabilities for the permanent disability model using the formulas derived above.

Example 5.4.1 In a permanent disability model there are three states: healthy, denoted by 0, disabled, denoted by 1, and dead, denoted by 2. Possible transitions are from 0 to 1, 0 to 2, and from 1 to 2. Suppose that the transition intensities for this model are specified as

(i) $\mu_x^{01} = 0.03$, $\mu_x^{02} = \mu_x^{12} = 0.025$, and

(ii) $\mu_x^{01} = a_1 + b_1 e^{c_1 x}$, $\mu_x^{02} = \mu_x^{12} = a_2 + b_2 e^{c_2 x}$, where, $a_1 = 5 \times 10^{-4}$, $b_1 = 3.5 \times 10^{-6}$, $c_1 = 0.14$, $a_2 = 6 \times 10^{-4}$, $b_2 = 7.5 \times 10^{-6}$, $c_2 = 0.09$.

For both the cases, compute the probability that (60) (i) remains in healthy state for next 10 years, (ii) becomes disabled in next 10 years, and (iii) dies in next 10 years.

Solution In the permanent disability model, neither state 0 nor state 1 can be reentered once it has been left; hence, for $i = 0, 1$, $_t p_x^{\overline{ii}} = {_t p_x^{ii}}$.

(i) Thus, under the first set up, the probability that (60) remains in healthy state for next t years is given by

$$_t p_{60}^{00} = {_t p_{60}^{\overline{00}}} = \exp\left\{-\int_0^t (0.03 + 0.025)\, ds\right\} = \exp\{-0.055t\},$$

which gives $_{10} p_{60}^{00} = \exp\{-0.55\} = 0.5769$. Thus, the chance that (60) remains in healthy state for next 10 years is 58 %.

Probability that (60) becomes disabled in next 10 years is denoted by $_{10} p_{60}^{01}$. Since neither state 0 nor state 1 can be reentered once it has been left, the formula to compute $_{10} p_{60}^{01}$ is given by

$$_{10} p_{60}^{01} = \int_0^{10} {_u p_{60}^{00}} \mu_{60+u}^{01} {_{10-u} p_{60+u}^{11}}\, du.$$

Intuitively this expression is interpreted as follows. For the member to move from state 0 to 1 between ages x to $x + t$, the member may remain in the same state 0 for some time, may be up to age $x + u$; then there is a transition from state 0 to state 1 between ages $x + u$ to $x + u + du$, where du is very small, with the chance of such an event given by $\mu_{x+u}^{01}\, du$; when transferred to state 1, the member remains in the same state up to age $x + t$. To compute $_{10} p_{60}^{01}$, first we need to compute $_{10-u} p_{60+u}^{11}$, which is given by

$$_{10-u} p_{60+u}^{11} = \exp\left\{-\int_0^{10-u} 0.025\, ds\right\} = \exp\{-0.025(10 - u)\}.$$

Hence,

$$_{10} p_{60}^{01} = \int_0^{10} {_u p_{60}^{00}} \mu_{60+u}^{01} {_{10-u} p_{60+u}^{11}}\, du$$

$$= \int_0^{10} \exp\{-0.055u\} \times 0.03 \times \exp\{-0.025(10 - u)\}\, du$$

$$= 0.03 \exp\{-0.25\} \int_0^{10} \exp\{-0.03u\}\, du$$

$$= 0.2019.$$

Further, $_{10}p_{60}^{02} = 1 - (_{10}p_{60}^{00} + _{10}p_{60}^{01}) = 1 - (0.5769 + 0.2019) = 0.2212$. Thus, the chance that (60) remains disabled for next 10 years is 20 % while dies in next 10 years is 22 %.

(ii) For the second setup,

$$_{t}p_{60}^{00} = _{t}p_{60}^{\overline{00}} = \exp\left\{-\int_{0}^{t}\left(\mu_{60+u}^{01} + \mu_{60+u}^{02}\right)du\right\}$$

$$= \exp\left\{-\left[(a_1 + a_2)t + \frac{b_1}{c_1}e^{60c_1}\left(e^{c_1t} - 1\right) + \frac{b_2}{c_2}e^{60c_2}\left(e^{c_2t} - 1\right)\right]\right\}.$$

Hence, $_{10}p_{60}^{00} = 0.6896$. On similar lines,

$$_{t}p_{60}^{11} = _{t}p_{x}^{\overline{11}} = \exp\left\{-\int_{0}^{t}\mu_{60+u}^{12}du\right\}$$

$$= \exp\left\{-\left(a_2 t + \frac{b_2}{c_2}e^{60c_2}\left(e^{c_2t} - 1\right)\right)\right\}.$$

With this expression, we can compute $_{10}p_{60}^{01}$ as follows:

$$_{10}p_{60}^{01} = \int_{0}^{10} {}_{u}p_{60}^{00}\mu_{60+u}^{01}\,{}_{10-u}p_{60+u}^{11}\,du.$$

However, in this setup it is not possible to integrate analytically, but we find the value numerically. Numerical integration gives $_{10}p_{60}^{01} = 0.2566$. Further, $_{10}p_{60}^{02} = 1 - (_{10}p_{60}^{00} + _{10}p_{60}^{01}) = 1 - (0.6896 + 0.2566) = 0.0538$. Thus, for the second setup, (60) remains healthy in next 10 years with chance of 69 %, becomes disabled in next 10 years with chance of 26 %, while dies in next 10 years with chance of 5 %.

The main aim to study such transition probabilities is the computation of premiums when the transitions among the states are modeled by a Markov process. As a first step, we generalize the definitions of insurance and annuity functions to a multiple state framework. There is no standard notation for these functions. Suppose that (x) is currently in state i. The actuarial present value of annuity of 1 per annum, payable continuously, while (x) is in some state j, which may be equal i, in future, is denoted by $\bar{a}_x^{ij}$ and is defined as follows. Denoting by I the indicator function, we have

$$\bar{a}_x^{ij} = E\left[\int_{0}^{\infty}e^{-\delta t}I\left(X(t) = j | X(0) = i\right)dt\right]$$

$$= \int_{0}^{\infty}e^{-\delta t}E\left[I\left(X(t) = j | X(0) = i\right)\right]dt$$

$$= \int_{0}^{\infty}e^{-\delta t}\,{}_{t}p_{x}^{ij}\,dt.$$

Thus, if (x), who is in state i at time 0, transits to state j at time t, then 1 unit will be payable at time t. Its value at time 0 is $e^{-\delta t}$.

On similar lines, the actuarial present value of discrete annuity due of 1 per annum, while (x) is in some state j, in future, denoted $\ddot{a}_x^{ij}$, is defined as follows:

$$\ddot{a}_x^{ij} = \sum_{k=0}^{\infty} v^k\, {}_k p_x^{ij}.$$

If the time period for transitions is restricted to, say n years, the actuarial present value of continuous and discrete annuity due of 1 per annum is defined as follows:

$$\bar{a}_{x:\overline{n}|}^{ij} = \int_0^n e^{-\delta t}\, {}_t p_x^{ij}\, dt \quad \text{and} \quad \ddot{a}_{x:\overline{n}|}^{ij} = \sum_{k=0}^{n-1} v^k\, {}_k p_x^{ij}.$$

For insurance benefits, the payment is usually conditional on making a transition. For example, a death benefit is payable on transition to a dead state. Suppose that a unit benefit is payable when transition is to state k, when currently the individual is in state i. Then the actuarial present value of the death benefit is given by

$$\bar{A}_x^{ik} = \int_0^{\infty} e^{-\delta t} \sum_{j\neq k} {}_t p_x^{ij}\, \mu_{x+t}^{jk}\, dt.$$

Similarly, if the policy is for limited period, then the actuarial present value of the death benefit in that limited period, denoted $\bar{A}_{x:\overline{n}|}^{ik}$, is given by

$$\bar{A}_{x:\overline{n}|}^{ik} = \int_0^n e^{-\delta t} \sum_{j\neq k} {}_t p_x^{ij}\, \mu_{x+t}^{jk}\, dt.$$

Premiums are then found using equivalence principle. We illustrate it for the permanent disability insurance model.

Example 5.4.2 In a permanent disability model, there are three states: healthy, denoted by 0, disabled, denoted by 1, and dead, denoted by 2. Possible transitions are from 0 to 1, 0 to 2, and from 1 to 2. Suppose that the transition intensities for this model are specified as, $\mu_x^{01} = 0.03$ and $\mu_x^{02} = \mu_x^{12} = 0.025$. An insurer issues 5-year permanent disability income insurance policy to a healthy life aged 45.

(i) Premiums are payable continuously while in the healthy state. The benefit of Rs 25000/- per year is payable continuously while in the disabled state, and the death benefit of Rs 100000/- is payable immediately on death. Calculate the premium when the force of interest is 5 % per year.
(ii) Suppose that premiums are payable yearly in advance while in the healthy state. The benefit of Rs 25000/- per year is payable yearly in arrears while in the disabled state, and the death benefit of Rs 100000/- is payable immediately on death. Calculate the premium when the force of interest is 5 % per year.

Solution We use the equivalence principle to find the premium. So we find the actuarial present value of inflow via premiums and actuarial present value of outflow via benefit payments.

(i) Let P denote the premium payable continuously per year while in healthy state in order to get the specified benefits. Then the actuarial present value of inflow via premiums is given by

$$P\bar{a}^{00}_{45:\overline{5}|} = P \int_0^5 e^{-\delta t} {}_t p^{00}_{45} \, dt.$$

For the given forces of transition,

$${}_t p^{00}_{45} = {}_t p^{\overline{00}}_{45} = \exp\left\{ -\int_0^t (0.03 + 0.025) \, ds \right\} = \exp\{-0.055t\}.$$

Hence,

$$P\bar{a}^{00}_{45:\overline{5}|} = P \int_0^5 e^{-0.05t} e^{-0.055t} \, dt = P \int_0^5 e^{-0.105t} \, dt = 3.889949 P.$$

The actuarial present value of outflow via disability benefit payments is given by

$$25000\bar{a}^{01}_{45:\overline{5}|} = 25000 \int_0^5 e^{-\delta t} {}_t p^{01}_{45} \, dt.$$

By definition,

$${}_t p^{01}_{45} = \int_0^t {}_u p^{00}_{45} \mu^{01}_{45+u} \, {}_{t-u} p^{11}_{45+u} \, du.$$

So first compute ${}_{t-u} p^{11}_{45+u}$, which is given by

$${}_{t-u} p^{11}_{45+u} = \exp\left\{ -\int_0^{t-u} 0.025 \, ds \right\} = \exp\{-0.025(t-u)\}.$$

Hence,

$$\begin{aligned}
{}_t p^{01}_{45} &= \int_0^t {}_u p^{00}_{45} \mu^{01}_{45+u} \, {}_{t-u} p^{11}_{45+u} \, du \\
&= \int_0^t \exp\{-0.055u\} \times 0.03 \times \exp\{-0.025(t-u)\} \, du \\
&= 0.03 \exp\{-0.025t\} \int_0^t \exp\{-0.03u\} \, du \\
&= \exp\{-0.025t\} - \exp\{-0.055t\}.
\end{aligned}$$

Thus,

$$\bar{a}^{01}_{45:\overline{5}|} = \int_0^5 e^{-0.05t} \left[e^{-0.025t} - e^{-0.055t} \right] dt = 0.2795273.$$

Hence,

$$25000\bar{a}^{01}_{45:\overline{5}|} = 25000 \times 0.2795273 = 6988.18.$$

The actuarial present value of outflow via death benefit payments is given by

$$100000\bar{A}^{02}_{45:\overline{5}|} = 100000 \int_0^5 e^{-\delta t} \left({}_tp^{00}_{45}\mu^{02}_{45+t} + {}_tp^{01}_{45}\mu^{12}_{45+t} \right) dt.$$

Now for the given forces of transition,

$$\bar{A}^{02}_{45:\overline{5}|} = \int_0^5 e^{-0.05t} \left(e^{-0.055t} + \left[e^{-0.025t} - e^{-0.055t} \right] \right) \times 0.025 \, dt = 0.1042369.$$

Hence,

$$100000\bar{A}^{02}_{45:\overline{5}|} = 10423.69.$$

By the equivalence principle, the premium P is given by

$$P = (25000 \times 0.2795273 + 100000 \times 0.1042369)/3.889949 = 4476.12.$$

(ii) In this case, the premiums are paid yearly in advance for 5 years while (45) is in healthy state, so it forms an annuity due. The benefit of Rs 25000/- per year is payable in arrears while in disabled state, so it forms an annuity immediately. The death benefit of Rs 100000/- is payable immediately on death. Here also we use the equivalence principle to find the premium. Let P denote the premium payable at the beginning of the year, per year in order to get the specified benefits while in healthy state. The actuarial present value of the premiums is given by

$$P\ddot{a}^{00}_{45:\overline{5}|} = P\sum_0^4 v^k {}_kp^{00}_{45} = P\sum_0^4 e^{-0.05k - 0.055k} = 4.097744P.$$

It is to be noted that $\ddot{a}^{00}_{45:\overline{5}|} = 4.097744 > \bar{a}^{00}_{45:\overline{5}|} = 3.889949$, as expected. The actuarial present value of outflow via disability benefit payments is given by $25000a^{01}_{45:\overline{5}|}$, where

$$a^{01}_{45:\overline{5}|} = \sum_1^5 v^k {}_kp^{01}_{45} = \sum_1^5 e^{-0.05} \left(e^{-0.025k} - e^{-0.055k} \right)$$

$$= \sum_1^5 \left(e^{-0.075k} - e^{-0.105k} \right) = 0.325775.$$

The actuarial present value of outflow via death benefit payments remains same as in (i). By the equivalence principle, the premium P is given by

$$P = (25000 \times 0.325775 + 100000 \times 0.1042369)/4.097744 = 4531.29.$$

We now proceed to discuss the application of Markov model to disability income insurance. In a disability income insurance model, there are three states: healthy, denoted by 0, sick, denoted by 1, and dead, denoted by 2. Possible transitions are from 0 to 1, 0 to 2, 1 to 0, and from 1 to 2. Disability income insurance policy pays a benefit during period of sickness, the benefit ceases on recovery. Here also premiums are payable when a person is in healthy state. In this model, states 0 and 1 both can be reentered. As a consequence, the transition probability $_t p_x^{01}$ is the sum of the probabilities of exactly one transition from 0 to 1, the probability of three transitions, 0 to 1, 1 to 0, and 0 to 1 again, five transitions, and so on. The following is a procedure to evaluate $_t p_x^{01}$ in such cases. Suppose that the member is in state i at age x and we want to find the probability that at age $x + t + h$, the member is in state j, where i and j are any two not necessarily distinct states. Thus, we wish to find $_{t+h} p_x^{ij}$, which is derived using following steps:

$$
_{t+h} p_x^{ij} = P\left[X(x+t+h) = j | X(x) = i\right]
$$

$$
= \sum_{k \in S} P\left[X(x+t+h) = j, X(x+t) = k | X(x) = i\right]
$$

$$
= \sum_{k \in S} P\left[X(x+t+h) = j | X(x+t) = k\right] P\left[X(x+t) = k | X(x) = i\right]
$$

$$
= \sum_{k \in S} {}_h p_{x+t}^{kj} {}_t p_x^{ik}
$$

$$
= {}_h p_{x+t}^{jj} {}_t p_x^{ij} + \sum_{k \in S, k \neq j} {}_h p_{x+t}^{kj} {}_t p_x^{ik}.
$$

We have assumed that for any positive interval of time h, the probability of two or more transitions within a time period of length h is $o(h)$. Hence, for small h,

$$
_h p_x^{ij} = h \mu_x^{ij} \quad \text{and} \quad _h p_x^{\overline{ii}} = {}_h p_x^{ii} = 1 - \sum_{k \neq i} {}_h p_x^{ik} = 1 - h \sum_{k \neq i} \mu_x^{ik}.
$$

Consequently, for small h,

$$
_h p_{x+t}^{jj} = {}_h p_{x+t}^{\overline{jj}} = 1 - \sum_{k \neq j} {}_h p_{x+t}^{jk} = 1 - \sum_{k \neq j} h \mu_{x+t}^{jk} \quad \text{and} \quad _h p_{x+t}^{kj} = h \mu_{x+t}^{kj}.
$$

Substituting these formulas into the last expression of $_{t+h} p_x^{ij}$, for small h, we get

$$
_{t+h} p_x^{ij} = {}_t p_x^{ij} \left(1 - \sum_{k \neq j} h \mu_{x+t}^{jk}\right) + \sum_{k \in S, k \neq j} h \mu_{x+t}^{kj} {}_t p_x^{ik}.
$$

This is a sort of recurrence relation to compute $_{t+h} p_x^{ij}$ from $_t p_x^{ik}$ for all k. Kolmogorov forward equations are derived from this identity. As an illustration, if we want to find $_{10}p_x^{ij}$, by choosing successively $t = 0, h, 2h, \ldots, 10-h$ with small step size h, we can use these formulas together with initial values $_0p_x^{ii} = 1$ and $_0p_x^{ij} = 0$ to calculate $_hp_x^{ii}, _hp_x^{ij}, _{2h}p_x^{ii}, _{2h}p_x^{ij}$, and so on until we get the values of $_tp_x^{ii}, _tp_x^{ij}$ for desired value of t. We then use these transition probabilities to compute the actuarial present values of premium and benefit and hence to compute the premium using the equivalence principle. The following example illustrates such a computation.

Example 5.4.3 In a disability income insurance model, there are three states: healthy, denoted by 0, sick, denoted by 1, and dead, denoted by 2. Possible transitions are from 0 to 1, 0 to 2, 1 to 0, and from 1 to 2. The transition intensities for the disability income insurance are as follows: $\mu_x^{01} = 0.002$, which is the force of transition for transition from healthy state to disabled state; $\mu_x^{10} = 0.1\mu_x^{01}$, which is the force of transition for transition from disabled state to healthy state; $\mu_x^{02} = \mu_x^{12} = A + BC^x$, where $A = 0.0007$, $B = 0.0001151$, $C = 1.096$. This gives the force of transition for transition from healthy state to dead state, which is same as the force of transition for transition from disabled state to dead state.

An insurer issues 5-year disability income insurance policy to a healthy life aged 45.

(i) Premiums are payable continuously while in the healthy state. The benefit of Rs 25000/- per year is payable continuously while in the disabled state, and the death benefit of Rs 100000/- is payable immediately on death. Calculate the premium when the effective rate of interest is 5 % per year.

(ii) Suppose that premiums are payable yearly in advance while in the healthy state. The benefit of Rs 25000/- per year is payable yearly in arrears while in the disabled state, and the death benefit of Rs 100000/- is payable immediately on death. Calculate the premium when the effective rate of interest is 5 % per year.

Solution We use the equivalence principle to find the premium. So we find the actuarial present value of inflow via premiums and actuarial present value of outflow via benefit payments.

(i) Let P denote the premium payable continuously per year while in healthy state in order to get the specified benefits. Then the actuarial present value of inflow via premiums is given by

$$P\bar{a}_{45:\overline{5}|}^{00} = P \int_0^5 e^{-\delta t} \, _tp_{45}^{00} \, dt.$$

The actuarial present value of outflow via disability benefit payments is given by

$$25000\bar{a}_{45:\overline{5}|}^{01} = 25000 \int_0^5 e^{-\delta t} \, _tp_{45}^{01} \, dt.$$

The actuarial present value of outflow via death benefit payments is given by

$$100000\bar{A}_{45:\overline{5}|}^{02} = 100000 \int_0^5 e^{-\delta t} \left(_tp_{45}^{00}\mu_{45+t}^{02} + _tp_{45}^{01}\mu_{45+t}^{12} \right) dt.$$

To obtain the transition probabilities, we proceed as follows. From the general expression for transition probabilities, for the disability income insurance model, we get

$$_{t+h}p_{45}^{00} = {_t}p_{45}^{00} - h\,{_t}p_{45}^{00}\left(\mu_{45+t}^{01} + \mu_{45+t}^{02}\right) + h\,{_t}p_{45}^{01}\mu_{45+t}^{10}$$

and

$$_{t+h}p_{45}^{01} = {_t}p_{45}^{01} - h\,{_t}p_{45}^{01}\left(\mu_{45+t}^{12} + \mu_{45+t}^{10}\right) + h\,{_t}p_{45}^{00}\mu_{45+t}^{01}.$$

By choosing successively $t = 0, h, 2h, \ldots, 5 - h$, we use these formulas together with initial values $_0p_x^{00} = 1$ and $_0p_x^{01} = 0$ to calculate $_h p_x^{00}, _h p_x^{01}, _{2h} p_x^{00}, _{2h} p_x^{01}$, and so on until we get the values of $_t p_x^{00}, _t p_x^{01}$ for $0 \leq t \leq 5$. An R code is given below to find these values recursively. We take $h = 1/200$.

```
x <- 45;
int <- 0.05;
del <- log(1+int);   del;
v <- 1/(1+int);   v;
a <- 0.0007;
b <- 0.0001151;
c <- 1.096;
m01 <- 0.002;
m10 <- 0.1*m01;
h <- 1/200
t <- seq(0, 5, h);
length(t);
m02 <- rep(0, 1001);
m12 <- rep(0, 1001);
for(i in 1:1001)
   {
   m02[i] <- a+(b*(c^(x+t[i])))
   m12[i] <- m02[i]
   }
v1 <- rep(0, 1000);
v2 <- rep(0, 1000);
p00 <- 1    #value of 0p45^00;
p01 <- 0    #value of 0p45^01;
p00 <- c(p00, v1);
p01 <- c(p01, v2);
for(i in 2:1001)
   {
   p00[i] <- p00[i-1]*(1-h*(m01+m02[i]))+p01[i-1]*h*m10;
   p01[i] <- p01[i-1]*(1-h*(m10+m12[i]))+p00[i-1]*h*m01;
   }
e <- exp(1);
length(p00);   length(p01);
```

```
t1 <- e^(-del*t);
a00 <- h*sum(t1*p00);
a00;
a01 <- h*sum(t1*p01);
a01;
abar <- h*sum(t1*((p00*m02)+(p01*m12)));
abar;
pcont <- (25000*a01+100000*abar)/a00;
pcont;
p00d <- c(p00[201], p00[401], p00[601], p00[801],
       p00[1001]);
p01d <- c(p01[201], p01[401], p01[601], p01[801],
       p01[1001]);
p00d;  p01d;
ad <- 1+v*p00[201]+v^2*p00[401]+v^3*p00[601]
     +v^4*p00[801];
ad;
ai <- v*p01[201]+v^2*p01[401]+v^3*p01[601]
     +v^4*p01[801]+v^5*p01[1001];
ai;
pdis <- (25000*ai+100000*abar)/ad;
pdis;
```

Using these R commands, we compute $_t p_{45}^{00}$ and $_t p_{45}^{01}$ for 1000 values of t with step size $1/200$ and hence the premiums.

The value of

$$\bar{a}_{45:\overline{5|}}^{00} = 4.326052$$

is found by using a Riemann sum approximation. The actuarial present value of outflow via disability benefit payments is given by

$$25000\bar{a}_{45:\overline{5|}}^{01} = 25000 \int_0^5 e^{-\delta t} {}_t p_{45}^{01} \, dt = 25000 \times 0.02060128.$$

The actuarial present value of outflow via death benefit payments is given by

$$100000\bar{A}_{45:\overline{5|}}^{02} = 100000 \int_0^5 e^{-\delta t} \left({}_t p_{45}^{00} \mu_{45+t}^{02} + {}_t p_{45}^{01} \mu_{45+t}^{12} \right) dt = 10^5 \times 0.04187389.$$

We use the equivalence principle to calculate the premium. Thus, the premium P is given by

$$P = (25000 \times 0.02060128 + 100000 \times 0.04187389)/4.326052 = 1087.00.$$

(ii) In this case the premiums are paid yearly in advance for 5 years while (45) is in healthy state, so it forms an annuity due. The benefit of Rs 25000/- per year is payable in arrears while in disabled state, so it forms an annuity immediate. The death benefit of Rs 100000/- is payable immediately on death. Here also we use the equivalence principle to find the premium. Let P denote the premium payable at the beginning of the year, per year in order to get the specified benefits while in healthy state. The actuarial present value of the premiums is given by

$$P\ddot{a}^{00}_{45:\overline{5}|} = P\sum_{0}^{4} v^k {}_k p^{00}_{45} = 4.453288\,P,$$

as from the recursive relations discussed in part (i) we get ${}_0 p^{00}_{45} = 1$, ${}_1 p^{00}_{45} = 0.9898919$, ${}_2 p^{00}_{45} = 0.9791849$, ${}_3 p^{00}_{45} = 0.9678342$, ${}_4 p^{00}_{45} = 0.9557928$. Further, $v = 0.952381$. It is to be noted that $\ddot{a}^{00}_{45:\overline{5}|} = 4.453288 > \bar{a}^{00}_{45:\overline{5}|} = 4.326052$, as expected. The actuarial present value of outflow via disability benefit payments is given by

$$25000 a^{01}_{45:\overline{5}|} = 25000\sum_{1}^{5} v^k {}_k p^{01}_{45} = 25000 \times 0.02421274$$

with ${}_1 p^{01}_{45} = 0.001981658$, ${}_2 p^{01}_{45} = 0.003923985$, ${}_3 p^{01}_{45} = 0.005822997$, ${}_4 p^{01}_{45} = 0.007674320$, and ${}_5 p^{01}_{45} = 0.009473168$. The actuarial present value of outflow via death benefit payments remains the same as in (i). By the equivalence principle, the premium P is given by

$$P = (25000 \times 0.02421274 + 100000 \times 0.04187389)/4.453288 = 1076.22.$$

5.5 Exercises

5.1 Suppose that the transitions among states in a critical illness model are governed by a homogeneous Markov chain. Let the state space be $S = \{H, C, D\}$, where H denotes the healthy state, C denotes the critically ill state, and D denotes the dead state. Let the transition probability matrix be as given below. What are the probabilities of being in each of the states after 1, 2, and 3 years, when at the beginning of the year the individual is in an active state?

$$Q = \begin{array}{c} \\ 1 \\ 2 \\ 3 \end{array} \begin{array}{ccc} 1 & 2 & 3 \\ \left(\begin{array}{ccc} 0.92 & 0.05 & 0.03 \\ 0.00 & 0.76 & 0.24 \\ 0 & 0 & 1 \end{array} \right). \end{array}$$

5.2 The following is a three-year term insurance plan. Insureds may be in one of the three states at the beginning of each year: active (1), disabled (2), or dead (3). All insureds are initially active. Transitions among the states are modeled by a time-homogeneous Markov chain with the following transition probability matrix:

$$
\begin{array}{ccc}
 & 1 & 2 & 3 \\
Q = \begin{array}{c} 1 \\ 2 \\ 3 \end{array} & \left(\begin{array}{ccc} 0.90 & 0.05 & 0.05 \\ 0.10 & 0.60 & 0.30 \\ 0 & 0 & 1 \end{array} \right).
\end{array}
$$

The benefit of Rs 10000 is payable at the end of year of death. Premiums are paid at the beginning of each year when the individual is active. Insurers do not pay premiums when they are disabled. Calculate the level annual net premium for this insurance when the effective annual rate of interest is 6 %.

5.3 For the CCRC model with transition probability matrices and matrices of cash flows as in Example 5.3.3, with annual effective interest rate 5 %,

   (i) Find the actuarial present value at time 4 of the cash flow resulting from the future transitions from state 2 to state 3, when the resident is in state 1 at time 4.

   (ii) Suppose that the resident is in state 1 at time 4. Find the benefit premium payable at the start of each future period in which the resident is in state 2, for the future transitions from state 2 to 3.

5.4 In a permanent disability model, there are three states: healthy, denoted by 0, disabled, denoted by 1, and dead, denoted by 2. Possible transitions are from 0 to 1, 0 to 2, and from 1 to 2. Let the transition intensities for this model be specified as $\mu_x^{01} = a_1 + b_1 e^{c_1 x}$ and $\mu_x^{02} = \mu_x^{12} = a_2 + b_2 e^{c_2 x}$, where $a_1 = 5 \times 10^{-4}$, $b_1 = 3.5 \times 10^{-6}$, $c_1 = 0.14$, $a_2 = 6 \times 10^{-4}$, $b_2 = 7.5 \times 10^{-6}$, $c_2 = 0.09$. An insurer issues 5-year disability income insurance policy to a healthy life aged 45.

   (i) Premiums are payable continuously while in the healthy state. The benefit of Rs 25000/- per year is payable continuously while in the disabled state, and the death benefit of Rs 100000/- is payable immediately on death. Calculate the premium when the effective rate of interest is 5 % per year.

   (ii) Suppose that premiums are payable yearly in advance while in the healthy state. The benefit of Rs 25000/- per year is payable yearly in arrears while in the disabled state, and the death benefit of Rs 100000/- is payable immediately on death. Calculate the premium when the effective rate of interest is 5 % per year.

5.5 Let the transition intensities for a disability income insurance model be specified as $\mu_x^{01} = a_1 + b_1 e^{c_1 x}$, $\mu_x^{10} = 0.1 \mu_x^{01}$, $\mu_x^{02} = \mu_x^{12} = a_2 + b_2 e^{c_2 x}$, where $a_1 = 5 \times 10^{-4}$, $b_1 = 3.5 \times 10^{-6}$, $c_1 = 0.14$, $a_2 = 6 \times 10^{-4}$, $b_2 = 7.5 \times 10^{-6}$, $c_2 = 0.09$. An insurer issues 5-year disability income insurance policy to a healthy life aged 50.

   (i) Premiums are payable continuously while in the healthy state. The benefit of Rs 25000/- per year is payable continuously while in the disabled state, and the death benefit of Rs 100000/- is payable immediately on death. Calculate the premium when the effective rate of interest is 5 % per year.

(ii) Suppose that premiums are payable monthly in advance while in the healthy state. The benefit of Rs 25000/- per year is payable monthly in arrears while in the disabled state, and the death benefit of Rs 100000/- is payable immediately on death. Calculate the premium when the effective rate of interest is 5 % per year.

Chapter 6
Stochastic Interest Rate

6.1 Introduction

In all the previous chapters, we have assumed that the rate of interest in the calculations of actuarial present values is fixed or deterministic and usually constant over
the period of policy. However, it is well accepted that the assumption of deterministic interest will be rarely realized in practice, particularly for long-term policies.
The simplest approach is to take into account the interest rates which vary with time
over the time period for which the policy is in force. In some problems in Chap. 5,
we have taken into account time-varying rates. It of course requires the projection
of interest rates over time, reflecting the economic status of society or of the region
under study. The other approach is to adopt the appropriate stochastic model for the
interest rates and find the actuarial present values, when time to decrement, mode
of decrement, and the interest rate also are random variables. The suitable model
for interest rates will depend on the variety of characteristics of capital market and
its behavior governed by many uncontrollable random factors. To build a stochastic model for the interest rate, a thorough study of the market conditions over long
periods is necessary.

In this chapter we discuss how the randomness in the interest rates is captured and
how it affects the actuarial present values of cash flows. In Sect. 6.2, we discuss how
different scenarios of time-varying interest rate are modeled by a random variable.
In Sect. 6.3, we discuss the commonly adopted stochastic model for the interest rate
and see how to obtain the actuarial present values of cash flows if the interest rates
for the period under study are assumed to be independent and identically distributed
random variables. In Sect. 6.4, the assumption of independence is relaxed, and interest rates over the time period are modeled by an appropriate time series model.

6.2 Random Interest Scenario

In this setup, interest rates are prescribed for each period for the time period for
which the cash flows are to be studied. Suppose that m such scenarios are visualized with a probability attached to each scenario. The probability distribution of the

S. Deshmukh, *Multiple Decrement Models in Insurance*,
DOI 10.1007/978-81-322-0659-0_6, © Springer India 2012

scenario reflects the chances of realization of the scenario depending on the various economic and market fluctuations. More precisely, let $i^{(l)} = \{i_r^{(l)}, r \geq 1\}$ denote the fixed rates for the time period under study for the lth scenario, $l = 1, 2, \ldots, m$, $i_r^{(l)}$ being the rate of interest for the rth transaction period. Let pr_l denote the probability assigned to the lth scenario. The discount factors corresponding to lth scenario are defined as follows:

$$v_0^{(l)} = 1, \qquad v_k^{(l)} = \prod_{r=1}^{k} (1 + i_r^{(l)})^{-1}, \quad k \geq 1.$$

Let $C^{(l)}$ denote the actuarial present value of the cash flow under interest scenario l. Then the actuarial present value which takes into account randomness of the interest scenarios is denoted by $_*C$ and is given by

$$_*C = \sum_{l=1}^{m} C^{(l)} pr_l.$$

For example, let $A_x(l)$ denote the actuarial present value of the unit benefit payable at the end of year of death in a whole life insurance issued to (x) when the interest scenario is specified by $i^{(l)} = \{i_r^{(l)}, r \geq 1\}$. Then

$$A_x(l) = E\left(v_{K+1}^{(l)}\right) = \sum_{k \geq 0} v_{k+1}^{(l)} \, {}_kp_x q_{x+k}, \quad \text{and} \quad _*A_x = \sum_{l=1}^{m} A_x(l) pr_l$$

is the actuarial present value of the death benefit under random scenario. Similarly, let $\ddot{a}_x(l)$ denote the actuarial present value of the unit benefit payable at the beginning of each year in a whole life annuity issued to (x) when the interest scenario is specified by $i^{(l)}$. Then

$$\ddot{a}_x(l) = \sum_{k \geq 0} v_k^{(l)} \, {}_kp_x \quad \text{and} \quad _*\ddot{a}_x = \sum_{l=1}^{m} \ddot{a}_x(l) pr_l.$$

By the equivalence principle, the premium, payable as discrete whole life annuity due, for a unit death benefit in a whole life insurance is then given by

$$_*P_x = \frac{_*A_x}{_*\ddot{a}_x}.$$

On similar lines, the benefit premium, payable as discrete n-year temporary annuity due, for a unit death benefit in an n-year term life insurance is given by

$$_*P_{x:\overline{n}|}^1 = \frac{_*A_{x:\overline{n}|}^1}{_*\ddot{a}_{x:\overline{n}|}} = \frac{\sum_{l=1}^{m} [\sum_{k=0}^{n-1} v_{k+1}^{(l)} \, {}_kp_x q_{x+k}] pr_l}{\sum_{l=1}^{m} [\sum_{k=0}^{n-1} v_k^{(l)} \, {}_kp_x] pr_l}.$$

The actuarial present values of benefit functions and annuity functions in single and multiple decrement models and also in multi-state transition models are

Table 6.1 Interest rate scenario

l	1	2	3	4	5	Probability
1	0.050	0.050	0.050	0.050	0.050	0.30
2	0.050	0.055	0.060	0.062	0.066	0.25
3	0.050	0.047	0.045	0.044	0.042	0.25
4	0.050	0.053	0.047	0.049	0.052	0.20

obtained on similar lines. We illustrate with the example how the random interest scenario is incorporated to compute the premiums.

Example 6.2.1 It is given that $q_{28} = 0.135$, $q_{29} = 0.146$, $q_{30} = 0.159$, $q_{31} = 0.173$, and $q_{32} = 0.188$. Find the annual premium paid as the 5-year temporary discrete annuity due, for the benefit of 1000, payable at the end of year of death, in a 5-year term insurance, issued to (28) under the random interest scenario as depicted in Table 6.1, where the ith row specifies the interest rates for the period of 5 years, $i = 1, \ldots, 4$. The last column specifies the probabilities attached to four scenarios.

Solution We use the formula

$$*P^1_{28:\overline{5}|} = \frac{*A^1_{28:\overline{5}|}}{*\ddot{a}_{28:\overline{5}|}} = \frac{\sum_{l=1}^{4}[\sum_{k=0}^{4} v^{(l)}_{k+1} \, {}_kp_{28}q_{28+k}]pr_l}{\sum_{l=1}^{4}[\sum_{k=0}^{4} v^{(l)}_{k} \, {}_kp_{28}]pr_l}.$$

We first compute the actuarial present value of benefit and premium for each scenario and using the probability structure for the random scenario find the premium for the 5-year term insurance issued to (28). The following set of R commands computes the actuarial present values and the corresponding premium:

```
i <- 0.05;
v <- (1+i)^(-1);
q <- c(0.135, 0.146, 0.159, 0.173, 0.188);
p <- 1-q;
p1 <- c(q[1], p[1]*q[2], p[1]*p[2]*q[3],
        p[1]*p[2]*p[3]*q[4], p[1]*p[2]*p[3]*p[4]*q[5]);
p2 <- c(1, p[1], p[1]*p[2], p[1]*p[2]*p[3],
        p[1]*p[2]*p[3]*p[4]);
v1 <- c(1, v, v^2, v^3, v^4, v^5);
at1 <- sum(p1*v1[2:6]);
ad1 <- sum(p2*v1[1:5]);
i2 <- c(0.050, 0.055, 0.060, 0.062, 0.066);
j2 <- 1/(1+i2);
v2 <- 1:6;
for (i in 2:6)
  {
  v2[i] <- v2[i-1]*j2[i-1]
  }
```

Table 6.2 Actuarial present values

Scenario l	Probability	$A^1_{28:\overline{5}|}$	$\ddot{a}_{28:\overline{5}|}$
1	0.30	0.50868	3.45319
2	0.25	0.50145	3.43175
3	0.25	0.51259	3.46491
4	0.20	0.50837	3.45169

```
at2 <- sum(p1*v2[2:6]);
ad2 <- sum(p2*v2[1:5]);
i3 <- c(0.050, 0.047, 0.045, 0.044, 0.042);
j3 <- 1/(1+i3);
v3 <- 1:6;
for (i in 2:6)
  {
  v3[i] <- v3[i-1]*j3[i-1]
  }
at3 <- sum(p1*v3[2:6]);
ad3 <- sum(p2*v3[1:5]);
i4 <- c(0.050, 0.053, 0.047, 0.049, 0.052);
j4 <- 1/(1+i4);
v4 <- 1:6;
for (i in 2:6)
  {
  v4[i] <- v4[i-1]*j4[i-1]
  }
at4 <- sum(p1*v4[2:6]);
ad4 <- sum(p2*v4[1:5]);
at <- c(at1, at2, at3, at4);
ad <- c(ad1, ad2, ad3, ad4);
pr <- c(0.30, 0.25, 0.25, 0.20);
l <- 1:4;
d <- data.frame(l, pr, at, ad);
d;
prm <- 1000*sum(at*pr)/sum(ad*pr);
prm;
```

The premium for 1000 units benefit is 147.17. Table 6.2 reports the actuarial present values corresponding to each scenario.

If the interest rate remains the same for all the five years, then the premium is $1000 \times 0.50868/3.45319 = 147.31$. For the given setup of random interest scenario, the two premiums are close to each other.

From the example it is clear that the important component in such a setup is modeling the interest scenario for the time period for which the policy is in force and assigning appropriate probability weightage for each scenario.

In the next section, we discuss a parametric model for the interest rate and assume that the interest rates over the periods are independent and identically distributed random variables.

6.3 Parametric Models

Let I_k denote the random effective interest rate for the kth transaction period. In the deterministic interest setup, I_k is a degenerate random variable, degenerate at i. The random variable corresponding to the present value at the beginning of first transaction period of one unit payable at the end of the nth transaction period, that is, the discount factor random variable, denoted V_n, is defined as

$$V_n = \prod_{k=1}^{n} (1 + I_k)^{-1}.$$

If $I_k, k = 1, \ldots, n$ are degenerate at i, then $V_n = v^n$, where $v = (1 + i)^{-1}$.
   It follows from Jensen's inequality that

$$E(1 + I_k)^{-1} \geq \left(1 + E(I_k)\right)^{-1}.$$

It is to be noted that $(1 + I_k)^{-1}$ is the random variable corresponding to the present value at the beginning of the kth transaction period of one unit payable at the end of the kth transaction period. Thus, the expected present value of 1 unit paid at the end of one period, in random setup cannot be less than the present value of the payment at the expected interest rate.
   Our aim is to compute the actuarial present values and premiums under the random interest setup. Suppose that $\{I_k, k \geq 1\}$ are independent and identically distributed random variables distributed as I. The straightforward approach is to view all the actuarial present values of cash flows studied in the previous chapters as the conditional expectations, conditional on I_k, and then the expectation of these with respect to the distribution of I_k will produce the actuarial present values in random interest rate setup. For example, denoting, as in the previous section, by $_*A_x$ the actuarial present value of a unit benefit payable at the end of year of death in the whole life insurance, we have

$$_*A_x = E_I\big\{E_{K|I}(V_{K+1}|I)\big\}.$$

If we further assume that the curtate future lifetime random variable K and I are independent, then we get

$$_*A_x = E_I\big\{E_{K|I}(V_{K+1}|I)\big\} = E_I\bigg\{\sum_{k\geq 0} V_{k+1} \, {}_k p_x q_{x+k}\bigg\}.$$

It is to be noted that the assumption of independence of K and I is reasonably valid as randomness in K is governed by the mortality pattern while randomness in I

is governed by the economic conditions. We can compute the expectation once we identify the appropriate model for the interest rate. We discuss the following commonly adopted model. Suppose that the interest random variable I_k is modeled as

$$\log(1 + I_k) = \delta + \epsilon_k, \quad k = 1, 2, \ldots,$$

where δ is a nonnegative constant, and ϵ_k are independent and identically distributed random variables, known as random shocks. If the random shocks are degenerate at 0, then $\log(1 + I_k)$ is degenerate at δ, which matches with the relation between fixed effective rate of interest i and the force of interest δ. In fact, the relation $\log(1 + i) = \delta$ leads to the model $\log(1 + I_k) = \delta + \epsilon_k$. One more interesting feature of this model is that with log transformation it becomes a linear model in parameter δ. This model is known as a long-term mean force of interest subject to random shocks.

Once a suitable model is found for random shocks, the distribution of I_k or corresponding discount factor V_k or the random force of interest $\Delta_k = \log(1 + I_k)$ can be found. Frequently, ϵ_k are assumed to have the $N(0, \sigma^2)$ distribution. This assumption is also well justified by the celebrated Central Limit Theorem in view of the fact that ϵ_k can be visualized as an additive effect of many random factors which affect interest rate. With ϵ_k having a normal distribution, entire further mathematics becomes simple. Firstly, $\log(1 + I_k)$ has the $N(\delta, \sigma^2)$ distribution, implies that $(1 + I_k)$ has the lognormal distribution with parameters δ and σ^2. From the first two moments of lognormal distribution we get

$$E(1 + I_k) = \exp\{\delta + \sigma^2/2\} \quad \text{and} \quad Var(1 + I_k) = \left(e^{\sigma^2} - 1\right)e^{\{2\delta + \sigma^2\}}.$$

From the lognormal distribution of $(1 + I_k)$ we can find the distribution of the discount factor random variable V_n as follows. We have

$$V_n = \prod_{k=1}^{n}(1 + I_k)^{-1} \quad \Leftrightarrow \quad \log V_n = -\sum_{k=1}^{n}\log(1 + I_k).$$

Under the assumption that I_k, $k = 1, 2, \ldots, n$ are independent and identically distributed random variables with $\log(1 + I_k)$ having the $N(\delta, \sigma^2)$ distribution,

$$\log V_n = -\sum_{k=1}^{n}\log(1 + I_k) \quad \text{has the } N\left(-n\delta, n\sigma^2\right) \text{ distribution,}$$

and hence V_n has the lognormal distribution with parameters $-n\delta$ and $n\sigma^2$. Using the formula for mean and variance of the lognormal distribution, we get

$$E(V_n) = e^{-n(\delta - \sigma^2/2)} \quad \text{and} \quad Var(V_n) = \left(e^{n\sigma^2} - 1\right)\left(e^{n(-2\delta + \sigma^2)}\right).$$

As a consequence, we have

$$*A_x = E_V\left\{\sum_{k\geq 0}V_{k+1}\, {}_kp_xq_{x+k}\right\} = \sum_{k\geq 0}e^{-(k+1)(\delta - \sigma^2/2)}\, {}_kp_xq_{x+k} = A_x\left(\delta'\right),$$

where $\delta' = \delta - \sigma^2/2$. Thus, $_*A_x$ is A_x calculated at the fixed force of interest δ'. On similar lines, we compute the annuity functions. The actuarial present value of the whole life annuity with unit benefit issued to (x) in this setup is given by

$$_*\ddot{a}_x = E_I\left(\sum_{k\geq 0} V_k \,_kp_x\right) = \sum_{k\geq 0} e^{-k(\delta-\sigma^2/2)} \,_kp_x = \ddot{a}_x(\delta').$$

Thus, for the annuity function, also $_*\ddot{a}_x$ is $\ddot{a}_x$ calculated at the fixed force of interest δ'. By the equivalence principle, the premium payable as a discrete life annuity due for the unit benefit in the whole life insurance payable at the end of year of death is given by

$$_*P_x = \frac{_*A_x}{_*\ddot{a}_x}.$$

The actuarial present values of benefit functions and annuity functions in single and multiple decrement models and also in multi-state transition models are obtained on similar lines, that is, first by conditioning on the interest rate and then taking the expectation with respect to the distribution of the interest rate. With the lognormal distribution of V_n, we get explicit expressions for many actuarial present values. This may not be true for any arbitrary distribution of V_n.

Example 6.3.1 Suppose that interest rate random variable is modeled as $\log(1 + I_k) = \delta + \epsilon_k$, $k = 1, 2, \ldots$, where $\delta = 0.05$, and ϵ_k follows $N(0, 0.001)$. Suppose that the force of mortality follows Gompertz' law given by $\mu_x = BC^x$ with $B = 0.0001$ and $C = 1.098$.

(i) Find the premium payable as a whole life annuity due for the benefit of Rs 1000/- payable at the end of year of death in a whole life insurance issued to (25).

(ii) Find the same if the interest rate is deterministic with force of interest $\delta = 0.05$.

Solution With the Gompertz law for mortality, with $m = B/\log C$, we have

$$_kp_x = e^{-mC^x(C^k-1)} \quad \text{and} \quad _kp_x q_{x+k} = e^{-mC^x(C^k-1)} - e^{-mC^x(C^{k+1}-1)}.$$

We use the formula for the premium as derived above with $\delta' = 0.05 - 0.0005 = 0.0495$. The following set of R commands finds the premium in both the cases:

```
e <- exp(1);
b <- 0.0001;
a <- 1.098;
m <- b/log(a, base=exp(1));
del <- 0.05;
d <- 0.001   #σ²;
v <- e^(-(del-d/2));
x <- 25;
```

```
k <- 0:(100-x);
y25 <- e^(-m*a^x*(a^k-1))-e^(-m*a^x*(a^(k+1)-1))
                                                   # P[K(x)=k];
s25 <- e^(-m*a^x*(a^k-1))   # kpx;
ab <- sum(v^(k+1)*y25)   # *Ax;
ad <- sum(v^k*s25)   # *äx;
p <- 1000*ab/ad   # *Px;
ab;   ad;   p;
```

(i) We have

$$*A_x = 0.1479758, \qquad *\ddot{a}_x = 17.64214, \quad \text{and} \quad 1000 \, *P_x = 8.39.$$

(ii) We use the same set of commands with $\sigma^2 = d = 0$, to find premium in the deterministic interest rate with $\delta = 0.05$. We get

$$A_x = 0.1455162, \qquad \ddot{a}_x = 17.52048, \quad \text{and} \quad 1000P_x = 8.31.$$

Remark 6.3.1 For the given mortality pattern and random interest rate, we find that in Example 6.3.1, the premiums in the deterministic and random setups are quite close. Such a closeness is a consequence of a small value of variance of the random shocks. It is to be noted that value of variance of the random shocks cannot be very large in light of possible values of $\log(1 + I_k)$, which are between 0 and $\log_e 2 = 0.69316$ as I_k ranges from 0 to 1. Under the assumption of normality of random shocks, with 99 % probability, $-3\sigma \leq \epsilon_k \leq 3\sigma$, so that with 99 % probability, $-3\sigma + \delta \leq \log(1 + I_k) \leq 3\sigma + \delta$. A specified value of δ imposes conditions on the possible values of σ, which are usually very small.

In this section, the randomness of the interest rate is captured by the parametric models. With the normal distribution for the shock and with the assumption that the rate of interest are independent and identically distributed random variables for transaction periods under study, the actuarial present values of cash flows can be easily computed. However, in practice, the assumption of independence may not be acceptable in some cases. One can then adopt suitable time series models for the interest rate over the periods in which the policy is in force. In the following section, we discuss one time series model.

6.4   Time Series Models

Time series models such as autoregressive and moving average models are extensively applied in a variety of situations. let $X_k = \log(1 + I_k)$. Then the autoregressive model of order one, AR(1), for $\{X_k, k \geq 1\}$ is given by

$$X_k - \delta = \phi(X_{k-1} - \delta) + \epsilon_k,$$

where ϵ_k is white noise in the sense that ϵ_k, $k \geq 1$, have zero expectation and are uncorrelated with common variance σ^2. The moving average model of order one, MA(1), for $\{X_k, k \geq 1\}$ is given by

$$X_k - \delta = \epsilon_k - \theta\epsilon_{k-1}, \quad \text{where } \epsilon_k \text{ is white noise, and } \epsilon_0 \text{ is known.}$$

AR(1) and MA(1) models are combined to give

$$X_k - \delta = \phi(X_{k-1} - \delta) + \epsilon_k - \theta\epsilon_{k-1}, \quad \text{where } \epsilon_k \text{ is white noise, and } \epsilon_0 \text{ is known.}$$

AR(1) model for differences is given by

$$X_k - X_{k-1} = \phi(X_{k-1} - X_{k-2}) + \epsilon_k, \quad \text{where } \epsilon_k \text{ is white noise.}$$

In the following, we discuss MA(1) model given by $X_k - \delta = \epsilon_k - \theta\epsilon_{k-1}$. In this model, the force of interest has a long-term mean δ, and random economic shocks create deviations from mean. The shock ϵ_{k-1} for period $k-1$ has a delayed and moderated impact on the force of interest in period k of size $-\theta\epsilon_{k-1}$. We assume that ϵ_k, $k \geq 1$, are independent and each has $N(0, \sigma^2)$ distribution. We further assume that $|\theta| \leq 1$, so that the model is stationary. Under this model, with $X_k = \log(1 + I_k)$, we have

$$(1 + I_k) = e^{X_k} = e^{\delta + \epsilon_k - \theta\epsilon_{k-1}};$$

hence, the discount factor random variable is given by

$$V_n = \prod_{k=1}^{n}(1 + I_k)^{-1} = e^{-\sum_{k=1}^{n}(\delta + \epsilon_k - \theta\epsilon_{k-1})}.$$

Hence,

$$\log V_n = -\sum_{k=1}^{n}(\delta + \epsilon_k - \theta\epsilon_{k-1}) = -n\delta - \epsilon_n + \theta\epsilon_0 - (1 - \theta)\sum_{k=1}^{n-1}\epsilon_k.$$

We have assumed that ϵ_k, $k = 1, 2, \ldots, n$, are independent and identically distributed random variables each having $N(0, \sigma^2)$. Hence, the moment-generating function $M(t)$ of ϵ_k is given by $M(t) = \exp(t^2\sigma^2/2)$. Further, it is known that the moment-generating function of the sum of n independent and identically distributed random variables is the nth power of the common moment-generating function. Using these results, we can find the expectation of V_n as follows:

$$E(V_n) = e^{-n\delta + \theta\epsilon_0} M(-1)\big(M(\theta - 1)\big)^{n-1}$$

$$= e^{-n\delta + \theta\epsilon_0 + n(\theta-1)^2\sigma^2/2} M(-1)\big(M(\theta - 1)\big)^{-1} = Ce^{-n\delta'},$$

$n = 1, 2, \ldots$, where $C = e^{\theta\epsilon_0} M(-1)(M(\theta - 1))^{-1} = e^{\theta\epsilon_0 + (1 - (1-\theta)^2)\sigma^2/2}$ and $\delta' = \delta - \log M(\theta - 1) = \delta - (\theta - 1)^2\sigma^2/2$. Once we have an expression for the expected value of V_n, we can compute the actuarial present values first by conditioning on the interest rate and then taking the expectation with respect to the distribution of the interest rate. Thus, we have

$$_*A_x = E_I\left\{\sum_{k\geq 0} V_{k+1}\ {}_kp_xq_{x+k}\right\} = C\sum_{k\geq 0} e^{-(k+1)\delta'}\ {}_kp_xq_{x+k} = CA_x(\delta').$$

On similar lines, $\ddot{a}_x$ is given by

$$_*\ddot{a}_x = E_I\left(\sum_{k\geq 0} V_k\ {}_kp_x\right) = C\sum_{k\geq 0} e^{-k\delta'}\ {}_kp_x = C\ddot{a}_x(\delta').$$

By the equivalence principle, the premium payable as a discrete life annuity due for the unit benefit payable at the end of year of death in the whole life insurance is given by

$$_*P_x = \frac{_*A_x}{_*\ddot{a}_x}.$$

Other actuarial present values can be obtained on similar lines. The following example illustrates the computation of premium under this model.

Example 6.4.1 Suppose that interest rate random variable is modeled by MA(1) as $\log(1 + I_k) = \delta + \epsilon_k - \theta\epsilon_{k-1}$, $k = 1, 2, \ldots$, where $\delta = 0.05$, $\theta = 0.6$, $\epsilon_0 = 0$, and ϵ_k follows $N(0, 0.001)$ distribution. Suppose that the force of mortality follows Gompertz' law given by $\mu_x = BC^x$ with $B = 0.0001$ and $C = 1.098$. Find the premium payable as a whole life annuity due for benefit of Rs 1000/- payable at the end of year of death in a whole life insurance issued to (25).

Solution As in Example 6.3.1, with the Gompertz law for mortality, with $m = B/\log C$, we have

$$_kp_x = e^{-mC^x(C^k-1)} \quad \text{and} \quad {}_kp_xq_{x+k} = e^{-mC^x(C^k-1)} - e^{-mC^x(C^{k+1}-1)}.$$

We use the formula for the premium as derived above with $\delta' = 0.05 + (0.6 - 1)^2 \times 0.0005 = 0.05008$. The following set of R commands computes the premium:

```
e <- exp(1);
b <- 0.0001;
a <- 1.098;
m <- b/log(a, base=exp(1));
del <- 0.05;
d <- 0.001    #σ²;
t <- 0.6    #θ=0.6;
v <- e^(-(del-(t-1)^2*d/2));
x <- 25;
k <- 0:(100-x);
```

Table 6.3 Interest rate scenario

l	1	2	3	4	5	Probability
1	0.060	0.060	0.060	0.060	0.060	0.35
2	0.060	0.065	0.066	0.068	0.069	0.20
3	0.060	0.057	0.055	0.054	0.052	0.20
4	0.060	0.063	0.057	0.059	0.054	0.25

```
y25 <- e^(-m*a^x*(a^k-1))-e^(-m*a^x*(a^(k+1)-1))
                                                # P[K(x)=k];
s25 <- e^(-m*a^x*(a^k-1))   #kpx;
ab <- sum(v^(k+1)*y25)   # *Ax;
ad <- sum(v^k*s25)   # *äx;
p <- 1000*ab/ad   # *Px;
p;
```

We have $1000 *P_x = 8.32$, again very close to that in (i) and (ii) of Example 6.3.1.

6.5 Exercises

6.1 It is given that $q_{38} = 0.0013$, $q_{39} = 0.0014$, $q_{40} = 0.0016$, $q_{41} = 0.0017$, and $q_{42} = 0.0019$. Find the annual premium paid as a 5-year temporary discrete annuity due, for the benefit of 1000, payable at the end of year of death, in a 5-year term insurance, issued to (38) under the random interest scenario as depicted in Table 6.3, where the ith row specifies the interest rates for the period of 5 years, $i = 1, \ldots, 4$. The last column specifies the probabilities attached to four scenarios.

6.2 Suppose that interest rate random variable is modeled as $\log(1 + I_k) = \delta + \epsilon_k$, $k = 1, 2, \ldots$, where $\delta = 0.06$, and ϵ_k follows $N(0, 0.002)$. Suppose that the force of mortality follows Makeham's law given by $\mu_x = A + BC^x$ with $A = 0.0007$, $B = 0.0001$, and $C = 1.098$.

   (i) Find the premium payable as a whole life annuity due for the benefit of Rs 1000/- payable at the end of year of death in a whole life insurance issued to (30).

   (ii) Find the same if the rate of interest is deterministic with the force of interest $\delta = 0.06$.

6.3 Suppose that the interest rate random variable is modeled as MA(1) as $\log(1 + I_k) = \delta + \epsilon_k - \theta\epsilon_{k-1}$, $k = 1, 2, \ldots$, where $\delta = 0.06$, $\theta = 0.5$, $\epsilon_0 = 0$, and ϵ_k follows $N(0, 0.002)$ distribution. Suppose that the force of mortality follows Makeham's law given by $\mu_x = A + BC^x$ with $A = 0.0007$, $B = 0.0001$, and $C = 1.098$. Find the premium payable as a whole life annuity due for the benefit of Rs 1000/- payable at the end of year of death in a whole life insurance issued to (30).

References

Anderson, A. W. (1985). *Pension mathematics for actuaries*. Needham: A. W. Anderson.

Borowiak, D. S. (2003). *Financial and actuarial statistics: an introduction*. New York: Dekker.

Bowers Jr., N. L., Gerber, H. U., Hickman, J. C., Jones, D. A., & Nesbitt, C. J. (1997). *Actuarial mathematics* (2nd ed.). Sahaumburg: The Society of Actuaries.

Deshmukh, S. R. (2009). *Actuarial statistics: an introduction using R*. Hyderabad, India: Universities Press.

Deshmukh, S. R., & Purohit, S. G. (2007). *Microarray data: statistical analysis using R*. Delhi: Narosa Publications.

Dickson, D. C. M., Hardy, M. R., & Waters, H. R. (2009). *Actuarial mathematics for life contingent risks*. New York: Cambridge University Press.

Ihaka, R., & Gentleman, R. (1996). A language for data analysis and graphics. *J. Comput. Graph. Stat.*, 5(3), 299–314.

Johnson, R. C. E., & Johnson, N. L. (1980). *Survival models and data analysis*. New York: Wiley.

Promislow, S. D. (2006). *Fundamentals of actuarial mathematics*. England: Wiley.

Purohit, S. G., Gore, S. D., & Deshmukh, S. R. (2008). *Statistics using R*. Delhi: Narosa Publications.

Verzani, J. (2005). *Using R for introductory statistics*. Florida: Chapman & Hall/CRC.

Index

Made in the USA
Middletown, DE
01 September 2024